THE LOSS PREVENT ̄
CERTIFICATIO ̄

CW00524523

List of Approved Fire and Security Products and Services

1999

A specifiers' guide

LPCB APPROVED FIRE AND SECURITY PRODUCTS AND SERVICES

FOREWORD

The Loss Prevention Certification Board (LPCB) approves fire and security products throughout the world. It is a company owned by The Loss Prevention Council (LPC), a leading international authority on loss prevention and control. The LPC is owned by the Association of British Insurers and Lloyd's, and aims to improve the protection of people and property through reduction of risk.

The *LPCB List of Approved Fire and Security Products and Services* is the essential reference for those having a responsibility for the design, specification and purchase of fire and security protection products and services. It is intended for use in conjunction with the rules and recommendations of the Association of British Insurers and Lloyd's for insurance purposes, and the requirements of national and international regulatory bodies, specifiers and other users.

The List comprises products and services which have been certificated or approved by LPCB to appropriate specifications, such as British Standards, European and International Standards, and LPC Standards (LPS's).

Products and services in the List have been examined to an appropriate standard and thereafter are monitored on a sampling basis for continuing conformance through periodic audit.

Except where indicated, all firms have been satisfactorily assessed by LPCB to ISO 9001 or 9002, the Quality Systems standard. Details of the certificated scope for each firm are given in the publication *LPCB Register of Companies Assessed to ISO 9000*.

Certification Marking

LPCB Product Certification Marking applies to products or services which are certificated to standards which meet the full technical requirements for certification purposes. The marking takes either of the following forms and is shown in the list next to each entry:

A   B

The Marks are also used on products or installations, either in association with the appropriate Standard Number, or with the Reference Number which appears in this List.

Where the certification marking scheme has been accredited by UKAS, The United Kingdom Accreditation Service, the form of marking **A**, above left can be used. In all other cases the mark **B**, above right is used. LPCB has applied to UKAS for an extension in the scope of accreditation to cover additional schemes, which will permit the UKAS accredited form of marking to appear on a larger range of products and services in future issues of this List.

LPCB APPROVED FIRE AND SECURITY PRODUCTS AND SERVICES

Approval Marking

LPCB Approval Marking takes the following form:

The Mark is used in conjunction with the LPC Reference Number which appears in this List. It is applied to those products and services which meet LPCB Approval requirements. Approval marking is discontinued in favour of certification marking as schemes are re-graded and approved products and services are re-examined to the revised technical requirements necessary for certification. Approval schemes will still continue to be introduced for new products and services for which no published standard exists. In such cases, a transition period to certification is allowed.

Information on the World Wide Web

The contents of this book are published on the LPCB pages of the LPC web site on: *http://www.lpc.co.uk/lpcb* and are updated on a regular basis. Updated CD Roms are also available from our sales office.

LPCB has continued to expand the number and scope of entries in this 1999 publication as the nature and extent of its operations expands in line with world demand.

LPCB policy on EC Directives and Local Regulations

We advise that it is the responsibility of the specifier, manufacturer, authority having jurisdiction or other responsible person or body, to ensure that products comply with local regulations or specifications, e.g. EC Directives, Building Regulations etc.

For further information on LPCB's services, contact the Sales and Marketing Engineer at:

The Loss Prevention Certification Board Limited
Melrose Avenue, Borehamwood, Hertfordshire WD6 2BJ, United Kingdom
Telephone: +44 (0)181 236 9600 • Fax: +44 (0)181 236 9601
E-mail: info@lpc.co.uk.
Web site: http://www.lpc.co.uk

CONTENTS

CONTENTS

INDEX

INDEX

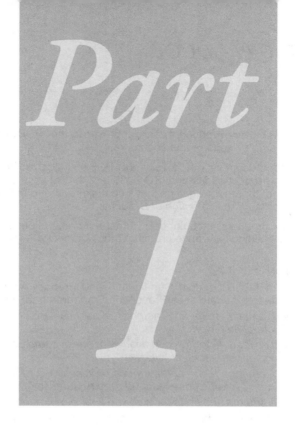

Part

1

BUILDING PRODUCTS

1 BUILDING PRODUCTS

INTRODUCTION

This Part contains building products that are approved by the LPCB and were commercially available at the time of issue of this document.

Certificated products comply with the relevant LPCB technical requirements. Certification is maintained by product audits and ISO 9001 or 2 surveillance undertaken by product experts. Listing of a product signifies that it complies with the requirements of the LPC *Design Guide for the Fire Protection of Buildings*.

The use of such products, in conjunction with the aforementioned document, can have an effect on the building insurance premium thus potentially reducing the service costs of buildings.

CERTIFICATION

Requirements for LPCB Product Certification and Product Approval are defined in ISO 9002 and the appropriate Loss Prevention Standards from the following:

LPS 1107 *Requirements, tests and methods of assessment of passive fire protection systems for structural steelwork.*

LPS 1108 *Quality schedule for the certification of passive and active (intumescent) fire protection products for structural steelwork.*

LPS 1124 *Requirements, tests, and methods of assessment for LPCB certification of active fire protection systems (intumescent) for structural steelwork.*

LPS 1132 *Requirements and tests for wall and floor penetration seals.*

LPS 1158 *Requirements and tests for fire resistant glazing, safety glazing and firescreens.*

LPS 1173 *Requirements for the certification of applicators of fire protection products for the protection of structural steel and other substrates.*

LPS 1181 *Requirements and tests for wall and ceiling lining materials and composite cladding materials.*

LPS 1195 *Requirements and tests for the approval of temporary buildings for use on construction sites.*

LPS 1207 *Fire requirements for protective covering materials.*

LPS 1208 *Fire performance requirements for metal-faced fire-resisting insulated panels.*

LPS 1215 *Flammability requirements for scaffold cladding materials.*

LPS 1220 *Test and performance requirements for passive fire protection systems for upgrading insulated panels*

LPS 1220 has been developed as the standard establishing fire resistance criteria for the upgrading of combustible core insulated composite panels. Combustible core panels protected with a listed product will have equivalent performance to systems listed in accordance with LPS 1208 : *Fire performance requirements for metal-faced fire resisting insulated panels.* Upgrading is accomplished by the application of a passive fire protection system to the face of insulated wall or ceiling panels.

Fire Protection Limited
Millars 3, Southmill Road, Bishop's Stortford, Hertfordshire CM23 3DH

Tel: +44 (0)1279 467 077 • Fax: +44 (0)1279 466 994

Installer of fire resistant ducts.

Rankins (Glass) Company Limited
The London Glass Centre, 24-34 Pearson Street, London E2 8JD

Tel: +44 (0)171 729 4200 • Fax: +44 (0)171 729 9197

Installer of glazed firescreens and security screens

Celotex Limited

Lady Lane Industrial Estate, Lady Lane, Hadleigh, Suffolk IP7 6BA

Sales Enquiries: Tel: +44 (0)181 579 0811 • Fax: +44 (0)181 579 0106

Certificate No. 093a to LPS 1181

LPCB Ref. No.	LPC Ref. No.	Trade Name	Specification	Support
093a/02	469	Celotex double-R Hydroform	20 - 100mm thick	Over & under purlin
093a/03	555	Celotex double-R Lining Grade LG2	20 - 100mm thick	Over & under purlin

Kingspan Industrial Insulation Limited

Blagden Chemical Site, Sully Moors Road, Sully, Penarth CF64 5RP.

Sales Enquiries: Tel: +44 (0)1457 853227/9, 861661 • Fax: +44 (0)1457 852319

Certificate No. 035a to LPS 1181

LPCB Ref. No.	LPC Ref. No.	Trade Name	Specification	Support
035a/02	-	K5	50mm thick	Over & under purlin

Kingspan Insulation Limited

Pembridge, Leominster, Herefordshire HR6 9LA

Tel: +44 (0)1544 387211 • Fax: +44 (0)1544 388888

Certificate No. 388a to LPS 1181

LPCB Ref. No.	Trade Name	Specification	Support
388a/01	Thermaliner	20 to 60 mm thick	Overpurlin

Rockwool Limited

Wern Tarw, Pencoed, Bridgend, Mid Glamorgan CF35 6NY

Tel: +44 (0)1656 862621 • Fax: +44 (0)1656 862302

LPC Ref. No.	Trade Name	Specification	Support
534	Rocklam lamella 2,4, 6, 8, 10, 11F, 12, 14,17 and 30	20-300mm thick (to nearest mm)	Overpurlin

For details of fire tested partition designs using Rocklam lamella slabs contact Technical Services Department, Rockwool.

Fire Resistance is the property of an element of building construction determined by test to British Standard 476: Parts 21-24: 1987 and defined in BS 476: Part 20: 1987; it is expressed in minutes.
There are three standards used within this section, LPS1181, LPS 1208 and BS 476: Part 21. The relevant standard is clearly shown below the telephone number of each company.

Isowall (UK) Limited
Unit 10, Alder Road, West Chirton Industrial Estate, North Shields, Tyne& Wear NE29 7TY

Tel: +44 (0)1912 585052 • Fax: +44 (0)1912 595742

Certificate No. 142a to LPS 1208

LPCB Ref. No.	Trade Name	Max (m)	Specification	Fire Resistance (min)		Type of Risk
				Integrity	Insulation	
142a/01	Isowall MF60W (Wall panel)	Height: 7.5	100mm thick	169	87	High
142a/02	Isowall MF120W (Wall panel)	Height: 7.5	100mm thick	120	120	High
142a/03	Isowall MF60C (Ceiling panel)	Span: 6.0	100mm thick	60	60	High

Kingspan Building Products Limited
Kingscourt, Co. Cavan, Ireland

Tel: +353 42 67172 • Fax: +353 42 98576

Certificate No. 186a to LPS 1181

LPCB Ref. No.	Trade Name	Specification	Fire Resistance (min)		Grade
			Integrity	Insulation	
186a/01	Kingspan KS1000RW/LPC (Wall panel)	40, 50, 60, 80 & 100mm thick	125	17	A
186a/02	Kingspan KS1000RW/LPC (Roof and wall panel)	30, 40, 50, 60, 80 & 100mm thick	-	-	B
186a/03	Kingspan KS1000BU/LPC (Roof and wall panel)	30, 40, 50, 60, 80 & 100mm thick	-	-	B

Note: All wall panels can be laid horizontally or vertically. For full details contact Kingspan.

Certificate No. 186b to LPS 1208

LPCB Ref. No.	Trade Name	Specification	Max. length m	Fire Resistance (min) BS 476 : Part 22		Type of Risk
				Int.	Ins.	
186b/01	Kingspan KS1000RW/LPC					
	Wall panel	100mm thick	Height 4.0	240	31	Normal
	Ceiling panel	100mm thick	Span 6.0	60	37	Normal

Notes: (a) Wall panels are laid vertically.
(b) Both wall and roof panels require internal steel flashings 0.7mm thick x 150mm wide stitched at 200mm centres to obtain fire resistance.

Kingspan Building Products Limited
Greenfield Business Park No. 2, Greenfield, Holywell, Clwyd CH8 7HU

Tel: +44 (0)1352 710111 • Fax: +44 (0)1352 712444

Certificate No. 260a to LPS 1181

LPCB Ref. No.	Trade Name	Specification	Fire Resistance (min)		Grade
			Integrity	Insulation	
260a/01	Kingspan KS600/900/1000 MR/LPC, MB/LPC & FL/LPC(Wall panel)	80 & 100mm thick	33	26	A
260a/02	Kingspan KS1000LP/LPC (Roof panel)	45, 50, 60, 80 & 100mm thick	-	-	B
260a/03	Kingspan KS600/900/1000/ 1200MR/LPC, MB/LPC & FL/LPC (Wall panel)	45, 50, 60, 80 & 100mm thick	-	-	B
260a/04	Kingspan KS1000CR/LPC (Roof panel)	45, 50, 60, 80 & 100mm thick	-	-	B

Note: Wall panels are laid horizontally.

Rockwool Limited
Wern Tarw, Pencoed, Bridgend, Mid Glamorgan CF35 6NY

Tel: +44 (0)1656 862621 • Fax: +44 (0)1656 862302

Approval to BS 476 : Part 21 (method 7) see note 1

LPCB Ref. No.	Trade Name	Specification	Fire Resistance (min)	
			Integrity	Insulation
022a/01	Rockwool Hardrock Roofing Board	(See note 2)	82	80

Notes:
1. The test construction was tested unloaded. However, the deflection of the roof did not exceed the allowable limits set out in BS†476†: Part†20†: 1987 for loadbearing horizontal elements.
2. The tested construction was made up as following:
   Sarnafil roof membrane
   2 layers of Rockwool Hardrock 180kg/m³ (40mm x 2)
   Monarflex reflex flame-proof vapour barrier
   Metal deck (nominally 0.7mm thick)
   203mm x 133mm x 30kg/m beams at 1680mm centres protected by 25mm thick Conlit 150.

Ward Building Components Limited

Sherburn, Malton, North Yorkshire YO17 8PQ

Tel: +44 (0)1944 710888 • Fax: +44 (0)1944 710555

Certificate No. 186a to LPS 1181
Certificate No. 260a to LPS 1181

LPCB Ref. No.	Trade Name	Specification	Fire Resistance (min)		Grade No.
			Integrity	Insulation	
186a/01	Moduclad IP1000 (Wall panel)	40, 50, 60, 80 & 100mm thick	120	17	A
186a/02	Moduclad IP1000 (Roof and wall panel)	30, 40, 50, 60, 80 & 100mm thick	-	-	B
260a/01	Moduclad DW1000 (Wall panel)	80mm thick	33	23	A
260a/02	Moduclad DW1000 (Roof Panel)	45, 60 and 80 mm thick	-	-	B
260a/03	Moduclad DW600/900/1000 (Wall panel)	45, 60 and 80 mm thick	-	-	B

Notes:(a) IP1000 reference 186a/01 requires internal joints to be stitched at 300mm centres to obtain fire resistance.
    (b) DW600/900/1000 reference 260a/01 and 260a/03 are laid horizontally.
    (c) DW1000 reference 260a/01 requires additional internal steel flashings 0.7mm thick x 150mm wide stitched at 250mm centres to obtain fire resistance.
For full details contact Ward.

1 SECTION 3:
FIRESCREENS AND PARTITIONS

NEW TEST STANDARD

The Loss Prevention Standard (LPS) 1158 has recently been rewritten by the Fire Resistant Glazing Panel in order to more closely meet the needs of the construction industry, regulators and users. The standard will be published at a similar time to this document and the first approvals will appear in the list amendments.

The insurers have been closely involved with the drafting of this standard and it will be referred to in the new Design Guide currently in preparation.

EAG Firemaster Limited

Unit D, Lea Industrial Estate, Lower Luton Road, Harpenden, Herts AL5 5EQ

Tel: +44 (0)1582 467111 • Fax: +44 (0)1582 769625

Approval to LPS 1158

LPC Ref. No.	Trade Name	Specification	Fire Resistance (min) Integrity	Insulation†
475	Solaglas Firemaster Profiled Steel Glazing System	5mm Firelite	90	0
		6mm Pilkington Georgian-wired	60	0
		6.4mm Guardian Villosa Georgian-wired	60	0
		6.5mm Pyran & 6mm Float (double-glazed)	45	0
		13.5mm Clear Firemaster	30	0
		6mm Clear Firemaster	45	0
477	Solaglas Firemaster Insulated Glazing System	19mm Pyrobel	90	60
478	Solaglas Firemaster Advanced Insulated Glazing System	20mm Pyrostop	60	60
		32mm Contraflam	60	60
		22mm Contraflam	45	30
479	Solaglas Firemaster Curtain Wall System	6.4mm Georgian-wired and 6mm Float (double-glazed)	60 (45*)	0
483	Solaglas Firemaster Series 90 Glazing System	5mm Firelite	210	0
		12mm Vegla Contrafeu 12	45	0
484	Solaglas Firemaster Horizontal Insulated System	20mm Pyrostop & 6mm Armourplate laminated	60	60

* when fitted as opening lights.
† where insulation is marked as 0 this denotes that the insulation was not measured or that it could not be measured in accordance with the relevant standard.

INSTALLERS
For installers see Part 1: Section 1

Integrity and insulation: properties determined in accordance with BS 476: Part 20: 1987 and expressed in minutes.

Hawke Cable Glands
Oxford Street West, Ashton-under-Lyne, Lancashire OL7 0NA

Tel: +44 (0)161 308 3611 • Fax: +44 (0)161 308 5848

Approval to LPS 1132

LPC Ref. No.	Trade Name	Use	Floor/ wall	Specification	Fire Resistance (min)	
					Integrity	Insulation
450	HC2,4,6,8 & multiples of	cables or steel pipes	both		120	120+
451	HCO2,4,6,8 & multiples of	cables or steel pipes	both		120	120+
452	HRT 70,100, 150,200	cables or steel pipes	both		240	120+
453	HRTO 70,100 150,200	cables or steel pipes	both		240	120+

NOTE: All 'X' versions of seals included
+ only if covered by a 100mm thick pad of mineral wool (110kg/m^3) on each side of the opening and the pipes lagged with a 50 mm thick mineral wool sheath for a length of 1m each side of the opening.

CAVITY FIRE STOPS

Rockwool Limited

Wern Tarw, Pencoed, Bridgend, Mid Glamorgan CF35 6NY

Tel: +44 (0)1656 862621 • Fax: +44 (0)1656 862302

Approval to LPS 1132

LPC Ref. No.	Trade Name	Specification	Fire Resistance (min)	
			Integrity	Insulation
535	Rockwool SP1	900mm long x 25 to 380mm thick x 150mm deep square ends	90	80
536	Rockwool SP2	900mm long x 25 to 380mm thick x 150 mm deep rebated ends	116	93
537	Rockwool SP3	900mm long x 25 to 380mm thick x 150mm deep rebated ends and 25mm thick Conlit 100 overlay	120	120

British Gypsum Limited

Fenton Lane, Sherburn-in-Elmet, Leeds, North Yorkshire LS25 6EZ

Tel: +44 (0)1917 682020 • Fax: +44 (0)1977 681289

Glasroc S has been approved jointly with the British Board of Agrément (BBA)
Approval to LPS 1107

LPC Ref.No.	Trade Name	Exposure	Additional Information
566	Glasroc S	Internal only	Thickness range 15 to 35mm Fire Resistance 30-120min (see table below)

H/A	Glasroc S Performance Table			
	30min	60min	90min	120min
30	15	15	20	25
50	15	15	20	25
70	15	15	20	25
90	15	15	20	25
110	15	15	20	25
130	15	15	20	30
150	15	15	25	35
170	15	15	25	35
190	15	15	25	35
210	15	15	25	35
230	15	15	25	35
250	15	20	30	35

Note: 1 The 30mm and 35mm thickness of protection can be in two separately fixed layers, i.e. 15mm + 15mm and 15mm + 20mm.

2 For protection thicknesses greater than 25mm mild steel supports must be provided. For two hours protection a mild steel angle board support must be provided. Soldiers can be used for board thickness not greater than 25mm and for periods of fire resistance up to 90 minutes.

Promat Fire Protection Limited
Victoria Works, Bonsall Street, Mill Hill, Blackburn, Lancashire BB2 4DD

Sales Enquiries: *Tel: +44 (0)1763 262310 • Fax: +44 (0)1763 262342*

Certificate No. 399a to LPS 1107

LPCB Ref. No.	Trade Name	Exposure	Additional information
399a/01	Vicuclad	Internal and sheltered external	Thickness range 16 to 80mm Fire resistance 30 to 240min (see table below)

Board thickness mm	Vicuclad Performance Table H_p/A					
	30min	60min	90min	120min	180min	240min
16	260	260	71	41	-	-
18	260	260	90	49	-	-
20	260	260	113	58	-	-
25	260	260	210	87	-	-
30	260	260	260	132	53	-
35	260	260	260	207	69	-
40	260	260	260	260	90	-
45	260	260	260	260	117	62
50	260	260	260	260	153	75
55	260	260	260	260	205	91
60	260	260	260	260	260	110
65	260	260	260	260	260	134
70	260	260	260	260	260	164
75	260	260	260	260	260	204
80	260	260	260	260	260	260

Notes: 1 Vicuclad 900 is used for 30min to 180min fire ratings, and Vicuclad 1050 for 240min fire ratings.

2 With box protection for beams, Vicuclad noggins the same thickness as the boards but not less than 25mm are fitted between the flanges for the attachment of the side boards at 610mm or 1000mm centres and the soffit board oversails these.

3 No noggins are required on beams or columns where 4-sided protection for up to 4 hours is required

4 Boards are bonded at all joints and to noggins with CV powder/liquid K cement or with Vicubond ready mixed cement. Nails are used to facilitate erection.

5 For two hour fire protection Vicuclad can be screw fixed without cement. Full details can be obtained from the manufacturer.

SECTION 7:
Protective Covering Materials

The following products comply with the regulations in *The Joint Code of Practice on the Protection from Fire of Construction Sites and Buildings Undergoing Renovation.*

British Sisalkraft
Commissioners Road, Stroud, Kent ME2 4ED

Tel: +44 (0)1634 292700 • Fax: +44 (0)1634 291029

Certificate No. 280a to LPS1207

LPCB Ref. No.	Trade Name	Thickness mm	Weight g/m²
280a/01	Seekure 830	0.16 to 0.25	96 to 150

Cito Kunststoffen BV
Industrieweg 12, 3881 LB Putten (Holland), Netherlands

UK Sales: Tel: +44 (0)800 834704 • Fax: +44 (0)800 834705

Certificate No. 364a to LPS 1207

LPCB Ref. No.	Trade Name	Thickness mm	Weight g/m²
364a/01	Tekgard	0.065 to 0.1 (before embossing) 0.3 to 0.35 (after embossing)	67 to 102

Correx Plastics
Madleaze Industrial Estate, Bristol Road, Gloucester GL1 5SG

Tel: +44 (0)1452 301893 • Fax: +44 (0)1452 300436

Certificate No. 227a to LPS 1207

LPCB Ref. No.	Trade Name	Thickness mm	Weight g/m²
227a/01	Correx Twinwall (Black)	2.2 ± 0.1	350 ± 4%
227a/02	Correx Twinwall (Coloured or translucent)	2.2 ± 0.1	350 ± 4%

IPB N.V.
Steenovenstraat 30, B-8790 Waregem, Belgium

UK Sales: Tel: +44 (0)800 834704 • Fax: +44 (0)800 834705

Certificate No. 365a to LPS 1207

LPCB Ref. No.	Trade Name	Thickness mm	Weight g/m²
365a/01	Proplex	2.0 to 3.0	250 to 400

Leonard Stace Limited
Gloucester Road, Cheltenham, Gloucestershire GL51 8NH

Tel: +44 (0)1242 514081 • Fax: +44 (0)1242 226422

Certificate No. 340a to LPS 1207

LPCB Ref. No.	Trade Name	Thickness mm	Weight g/m²
340a/01	Floortech	0.7 ± 15%	330 ± 5%

Manton Plastics Limited
Daniels Industrial Estate, Bath Road, Stroud, Gloucestershire GL5 3TL

Tel: +44 (0)1453 755700 • Fax: +44 (0)1453 755620

Certificate No. 324a to LPS 1207

LPCB Ref. No.	Trade Name	Specification	Thickness mm	Weight g/m²
324a/01	Frapol	LD	0.50 to 0.99	0.46 to 0.92
		MD	1.00 to 1.99	0.93 to 1.85
		HD	2.00 to 3.00	1.86 to 2.79

MegaFilm Limited
Factory J, Bone Lane Industrial Estate, Newbury, Berkshire RG14 5SH

Tel: +44 (0)1635 521494 • Fax: +44 (0)1635 521717

Certificate No. 226a to LPS 1207

LPCB Ref. No.	Trade Name	Specification
226a/01	MegaFilm	6LS-FR
		10LS-FR
		15LS-FR
		20LS-FR
		HLS Temporary Covering
		MLS Temporary Covering
		Megasheet

Nields Plastic Company Limited
Foley Works Estate, Hereford HR1 2SF

Tel: +44 (0)1432 268341 • Fax: +44 (0)1432 353247

Certificate No. 342a to LPS 1207

LPCB Ref. No.	Trade Name	Thickness mm	Weight g/m²
342a/01	Antimar	0.075 to 0.3	70 to 280±15%

Northern Ireland Plastics Limited
39 Shrigley Road, Killyleagh, Co. Down, Northern Ireland BT30 9SR

Tel: +44 (0)1396 828753/4 • Fax: +44 (0)1396 828809

Certificate No. 341a to LPS 1207

LPCB Ref. No.	Trade Name	Thickness mm	Weight g/m²
341a/01	Arrowtex	2.3 ± 0.2	200 to 800

Twinplast Limited
Unit 2, Greycaine Road, Watford, Hertfordshire WD2 4JP

Tel: +44 (0)1923 230191/817761 • Fax: +44 (0)1923 817756

Certificate No. 375a to LPS 1207

LPCB Ref No.	Trade Name	Thickness mm	Weight g/m²
375a/01	Twinflute (fire retardant)	2.0 to 4.0	250 to 1000
	Twinplast (fire retardant)		
	Antinox (fire retardant)		

TBA Textiles

P O Box 40, Rochdale, Lancashire OL12 7EQ

Tel: +44 (0)1706 647422 • Fax: +44 (0)1706 712283

Certificate No. 387a to BS 476: Part 22: 1987

LPCB Ref. No.	Trade Name	Specification	Fire Resistance (min)	
			Integrity	Insulation
387a/01	Firefly Plus 30	6.0 m long x 1.3 m wide	30	15
387a/02	Firefly Plus 60	6.0 m long x 1.3 m wide	60	15

Notes: (a)   Vertical joints between individual barriers are double-folded, incorporating 8mm staples at 100mm centres, (for full insulationinstructions see TBA textiles).

      (b)   These products have achieved the same fire ratings when penetrated by services detailed below.

           (i) Cable tray carrying   1 x 9mm, 2 core +EEC, FP200 cable.
                                          1 x 12mm, 2 core steel wire armoured cable.
                                          1 x 8mm, 4 core signal cable.
                                          1 x 7mm, Pyro 2L cable.
                                          1 x 6mm, 2 core Belden cable.

           (ii) 60mm diameter x 4.5mm wall thickness pipe.

Part

2

FIREBREAK DOORS AND SHUTTERS

2 *FIREBREAK DOORS AND SHUTTERS*

INTRODUCTION

This part contains a list of all doors and shutters that are approved by the LPCB and were commercially available at the date of issue of this document.

Doors and shutters are acceptable for both fire insurance and Building Regulations purposes (see *Approved Document B - Fire Spread;* HMSO).

Where Approved doors are to be installed, attention should be paid to the recommendations in the LPC *Design Guide for the Fire Protection of Buildings*. Approved fire doors/shutters may be recognised on site as each bears a metal plate quoting the relevant reference number. The approval is limited to the sizes shown on the list. In this connection, it should be noted that the sizes given refer to the dimensions of the wall opening, except in the case of steel rolling shutters, where they refer to the dimensions between the guides and between the sill and underside of the barrel enclosure. The acceptability of doors and shutters in excess of the quoted sizes is at the discretion of the authority concerned on an individual case basis.

As the effectiveness of a door in retaining capability is dependent upon its conditions of use, it is normally an insurance requirement that Approved fire doors and shutters shall be maintained in efficient working order and shall be kept closed except during working hours (whether or not automatic closing devices are fitted). It is also important that doors and shutters are kept free of goods that might ignite if subjected to radiant heat or might obstruct their operation and that their operation (including that of any automatic closing devices fitted) is regularly checked. When not in frequent use doors and shutters should be kept closed.

Assessments

Doors in excess of the maximum size that can be tested are assessed for satisfactory performance in terms of the BS 476 criteria.

Certification

Certification entails product test to the appropriate standards and assessment audit of the manufacturer's quality management system to ISO 9000. The assessment audit includes examination of installed doors and the procedure for the control of installation. Subsequent surveillance audits are conducted at six-monthly intervals, and include examination of a number of installed doors. The following standards apply:

LPS 1056 *Requirements and tests for fire doors, lift landing doors and shutters (including installation).*

Manufacturers seeking certification for their products should contact the LPCB and agree test programmes prior to testing. A representative of the LPCB witnesses the fire test.

The results of tests arranged without the agreement of the LPCB may not be acceptable.

Certificated products are identified in the following list by:

Maintenance

See section 6 of this part for maintenance and service companies. The companies in the list comply with ISO 9002 and:

LPS 1197 *Requirements for companies undertaking the maintenance and repair of doors and shutters.*

Compliance with LPS 1197 is identified in the following list by: **LPCB**

2 SECTION 1:
INSTALLATION REQUIREMENTS

Approved doors and shutters meet their designated fire resistance specifications when installed in accordance with these requirements.

The dimensions of all opening/doorways shall not exceed those given in the List. Sliding doors and folding shutters shall be in one leaf.

The sill and jambs of the opening shall be constructed of brick, reinforced concrete or concrete blocks having dense or light weight aggregate. Aerated concrete is not allowed. The sill, which may be covered by a steel plate, shall not be less than 100mm thick and shall extend throughout the opening and to not less than 75mm past each edge of the opening and from each face of the wall. It is recommended that the sill be raised at least 50mm to check the flow of extinguishing water and, in such cases, the sill should be ramped. No combustible floor or wall covering shall extend through the opening.

The head of the opening shall be of reinforced concrete at least 125mm deep having dense or lightweight aggregate with at least 50mm cover to the reinforcement or it may be a steel lintel provided it is protected by brickwork or concrete not less than 50mm thick.

In the case of rolling shutters, the barrel enclosure shall be supported on the jambs with a minimum bearing of 100mm at each end. Expansion clearances of at least 12.5mm per metre width shall be provided between the barrel enclosure and the enclosing jambs. Chases measuring nominally 115mm by 115mm shall be provided in the jambs to house the channel guides. The chases shall be set back at least 100mm from the front of the jamb. The barrel enclosure shall be housed completely within the wall opening.

In the case of other doors and shutters, no part of the leaf, frame or hanging assembly shall project more than 100mm from the face of the jambs and lintel when the door/shutter is in the closed position.

The wall opening shall be such as to provide flush contact with the door frame and a clearance between the sill and door/shutter, when closed, not in excess of 6mm. Where necessary, the opening may be rendered with cement mortar.

Where double Approved fire doors/shutters (i.e. two Approved fire doors/shutters of any type, one on each side of the wall) are to be installed, there shall be a gap between the inner faces of the doors/shutters, when closed, of not less than 225mm in the case of double rolling shutters or 150mm in other cases. In the case of double rolling shutters the barrels shall be installed at the same level and, if housed in the same enclosure, shall be separated by a dividing plate.

When the wall thickness is less than that required for a wall opening, the wall at the opening shall be built out using only materials as required above, such materials being bonded with,or tied into the wall, adequately supported, devoid of cavity and not less than 325mm in width (not less than 100mm in width where they enclose the ends of a rolling shutter barrel enclosure). The built-out head shall be constructed of reinforced concrete not less than 125mm thick and shall rest on, and extend to the full width and thickness of the built-out jambs.

The provision of a lobby between firebreak doors/shutters protecting a firebreak wall opening is permitted only if the walls enclosing the lobby are themselves constructed in accordance with the LPC *Design Guide for the Fire Protection of Buildings*. In the event of the floor or ceiling not complying with these requirements it is required that the enclosing walls extend through other floors until this condition is satisfied. All openings into such a firebreak lobby shall be protected by Approved fire doors/shutters.

Fire doors and shutters shall only be installed by trained personnel in accordance with the manufacturer's instructions. Expanding anchor bolt fixings into the wall opening shall be tightened to the manufacturer's recommended torques and shall not be through mortar joints.

Where there is a likelihood of waste accumulation, provision shall be made to prevent the fouling of any drop-bolt keep or bottom track. Waste deposits should be regularly cleared.

The fitting of devices to close the door/shutter automatically in the event of fire is recommended, but such automatic fire closing devices, or other self-closing devices, shall not interfere with the manual opening and closing of the door/shutter. Where such devices are fitted, it shall be ensured that double-leaf doors can close only in the correct order and that all bolts/latches engage in the frame on automatic closing.

Deviation from the above shall be covered by assessment or be part of the manufacturers certification. Appropriate documents shall be available to the certificated companies clients.

Controlled descent

It is recommended that fire doors and shutters have controlled closure where they are operated by automatic fire detection systems and occupants can use the opening for escape purposes.

2 SECTION 2:
FIREBREAK DOORS AND SHUTTERS

For details of the fire resistance required for different buildings see the LPC *Design Guide for the Fire Protection of Buildings.*
Where insulation is marked as 0 this denotes that the insulation was not measured or that it could not be measured in accordance with the relevant standard.

Accent-Hansen

Greengate Industrial Park, Greengate, Middleton, Manchester M24 1SW

Tel: +44 (0)161 284 4100 • Fax: +44 (0)161 655 3119

Certificate No. 021a to LPS 1056

LPCB Ref. No.	Spec.	Trade Name	Door Type	Maximum sizes			Fire resistance (min) BS 476 : Part 22	
				Height m	Width m	Area m²	integrity	insulation
021a/01	4	Fireshield	single-leaf hinged	2.74	1.22	2.6	120	0
021a/02	4	Fireshield	double-leaf hinged	2.74	2.44	5.2	120	0
021a/03	4	Fireshield	single-leaf hinged with	2.74	1.22	2.6		0
			6.5mm Pyran vision panel	0.45	0.35		90	
			6mm Georgian wired vision panel	0.30	0.30		60	0
021a/04	4	Fireshield	double-leaf hinged with	2.74	2.44	5.2		
			6.5mm Pyran vision panel	0.45	0.35		90	0

Amber Doors Limited

Mason Way, Platts Common Industrial Estate, Hoyland, Barnsley, S. Yorks S74 9TG

Tel: +44 (0)1226 351135 • Fax: +44 (0)1226 350176

Certificate No. 001a to LPS 1056

LPCB Ref. No.	Spec.	Trade Name	Door Type	Maximum sizes			Fire resistance (min) BS 476 : Part 22	
				Height m	Width m	Area m²	integrity	insulation
001a/01	4	Amber Rolling Shutter	face-fixed rolling shutter	7.00	7.00	49.0	240	0
001a/04	4	Protector	folding shutter	7.00	7.00	49.0	240	0
001a/05	4	Protector	single-leaf hinged	3.00	1.50	4.5	-	-
001a/06	4	Protector	double-leaf hinged	3.00	3.00	9.0	-	-

Ascot Industrial Doors Limited

Britannia Way Industrial Park, Union Road, Bolton BL2 2HE

Tel: +44 (0)990 556644 • Fax: +44 (0)1204 545800

Certificate No. 131a to LPS 1056

LPCB Ref. No.	Spec.	Trade Name	Door Type	Maximum sizes			Fire resistance (min) BS 476 : Part 22	
				Height m	Width m	Area m²	integrity	insulation
131a/01	4	A200 Series	rolling shutter	7.00	7.00	49.0	240	0

Attenborough Industrial Doors Limited
Merlin Way, Quarry Hill Industrial Estate, Ilkeston, Derbyshire DE7 4RA

Tel: +44 (0)115 930 0815 • Fax: +44 (0)115 944 8930

Certificate No. 145b to LPS1056

LPCB Ref. No.	Spec.	Trade Name	Door Type	Maximum sizes			Fire resistance (min) BS 476 : Part 22	
				Height m	Width m	Area m²	integrity	insulation
145b/01	4	Attenborough Fire Shutter	rolling shutter	7.00	7.00	49.0	240	0

Blount Shutters Limited
Unit B, 734 London Road, West Thurrock, Essex RM16 1NL

*Tel: +44 (0)1708 860000 • Fax: Service: +44 (0)1708 861271
Accounts/Sales: +44 (0)1708 861272*

Certificate No. 347a to LPS 1056

LPCB Ref. No.	Spec.	Trade Name	Door Type	Maximum sizes			Fire resistance (min) BS 476 : Part 22	
				Height m	Width m	Area m²	integrity	insulation
347/a01	4	Elite	rolling shutter	7.00	7.00	49.0	240	0

Bolton Gate Co Limited
Waterloo Street, Bolton, Lancashire BL1 2SP

Tel: +44 (0)1204 871000 • Fax: +44 (0)1204 871049

Certificate No. 026-2a to LPS 1056

LPCB Ref. No.	Spec.	Trade Name	Door Type	Maximum sizes			Fire resistance (min) BS 476 : Part 22	
				Height m	Width m	Area m²	integrity	insulation
026a/01	4	FireRoll E240	face-fixed rolling shutter	7.00	7.00	49.00	240	0
026a/02	4	FireRoll E240	built-in rolling shutter	3.66	4.27	15.6	240	0
026a/03	4	Escalator	rolling shutter	15.00	5.00	75.00	240	0
026a/04	4	Contour	vertically hung rolling shutter	4.50	15.00	67.5	240	0
026a/05	4	FireRoll E30	servery rolling shutter	3.0	3.0	9.0	30	0
				2.0	4.0	8.0	30	0
		FireRoll E60		3.0	3.0	9.0	60	0
				2.0	4.0	8.0	60	0
		FireRoll E120		3.0	3.0	9.0	120	0
				2.0	4.0	8.0	120	0

Chiltern Industrial Doors Limited
Unit 11, Commerce Way, Leighton Buzzard, Bedfordshire LU7 8RW

Tel: +44 (0)1525 383537 • Fax: +44 (0)1525 382314

Certificate No. 246c to LPS 1056

LPCB Ref. No.	Spec.	Trade Name	Door Type	Maximum sizes			Fire resistance (min) BS 476 : Part 22	
				Height m	Width m	Area m²	integrity	insulation
246c/01	4	Chiltern Fire Resisting Rolling Shutter	rolling shutter	7.00	7.00	49.0	240	0

Fortress Industries Limited

6 Trench Road, Newtownabbey, Co. Antrim, Northern Ireland BT36 8TY

Tel: +44 (0)1232 342655 • Fax: +44 (0)1232 342651

Certificate No. 146a to LPS 1056

LPCB Ref. No.	Spec.	Trade Name	Door Type	Maximum sizes			Fire resistance (min) BS 476 : Part 22	
				Height m	Width m	Area m²	integrity	insulation
146a/01	4	FF76	rolling shutter	7.00	7.00	49.0	240	0

Norman Hart (Newcastle) Limited

Redburn Road, Westerhope Industrial Estate, Newcastle upon Tyne NE5 1PJ

Tel: +44 (0)191 2140404 • Fax: +44 (0)191 2711611

Certificate No. 016a to LPS 1056

LPCB Ref. No.	Spec.	Trade Name	Door Type	Maximum sizes			Fire resistance (min) BS 476 : Part 22	
				Height m	Width m	Area m²	integrity	insulation
016a/01	4	Firebrand	face-fixed rolling shutter	7.00	7.00	49.0	240	0

Henderson-Bostwick Industrial Doors Limited

Grange Close, Clover Nook Industrial Estate, Somercotes, Derbyshire DE55 4QT

Tel: +44 (0)1773 523300 • Fax: +44 (0)1773 523301

Also at: Mersey Industrial Estate, Stockport, Cheshire SK4 3ED

Tel: +44 (0)161 947 4040 • Fax: +44 (0)161 947 4041

Certificate No. 002a to LPS 1056
Certificate No. 006a to LPS 1056

LPCB Ref. No.	Spec.	Trade Name	Door Type	Maximum sizes			Fire resistance (min) BS 476 : Part 22	
				Height m	Width m	Area m²	integrity	insulation
006a/08	4	Delta	folding shutter	7.00	7.00	49.0	240	0
002a/06	4	Defender	single-leaf hinged	3.00	1.50	4.5	240	0
002a/06	4	Fireguard	double-leaf hinged	3.00	3.00	9.0	240	0
006a/05	4	Firetex	face fixed rolling shutter	7.00	7.00	49.0	240	0
006a/09	4	Escalator	rolling shutter	15.00	5.00	75.00	240	0
006a/10	4	Sidewinder	vertically hung rolling shutter	4.50	15.00*	67.5	240	0

* Where this product is used in a firebreak wall insurers may wish to reduce the maximum size.

Industrial Door Engineering Limited

Winnington Avenue, Winnington, Northwich, Cheshire CW8 4EQ

Tel: +454 (0)1606 871832 • Fax: +44 (0)1606 871482

Certificate No. 165a to LPS 1056

LPCB Ref. No.	Spec.	Trade Name	Door Type	Maximum sizes			Fire resistance (min) BS 476 : Part 22	
				Height m	Width m	Area m²	integrity	insulation
165a/01	4	Sentry	rolling shutter	7.0	7.0	49.0	240	0
165a/02	4	Sentry 2	rolling shutter	7.0	7.0	49.0	120	0

LB Securities

Unit 6, Sterling Industrial Estate, Rainham Road South, Dagenham, Essex RM10 8TX

Tel: +44 (0)181 517 6655 • Fax: +44 (0)181 984 0378

Certificate No. 247a to LPS 1056

LPCB Ref. No.	Spec.	Trade Name	Door Type	Maximum sizes			Fire resistance (min) BS 476 : Part 22	
				Height m	Width m	Area m²	integrity	insulation
247a/01	4	Fireguard	rolling shutter	7.00	7.00	49.0	240	0

Lycetts (Burslem) Limited

Glendale Street, Burslem, Stoke on Trent ST6 2EP

Tel: +44 (0)1782 575236 • Fax: +44 (0)1782 577841

Certificate No. 029a to LPS 1056

LPCB Ref. No.	Spec.	Trade Name	Door Type	Maximum sizes			Fire resistance (min) BS 476 : Part 22	
				Height m	Width m	Area m²	integrity	insulation
029a/01	4	Lycetts 4 hour Face Fixed	rolling shutter	7.00	7.00	49.0	240	0
029a/02	3	Lycetts 4 hour built in	rolling shutter	3.66	4.27	15.6	240	0
029a/03	4	Lycetts 1 hour Face Fixed	rolling shutter	6.00	6.00	36.0	60	0

Mercian Shutters Limited

Pearsall Drive, Brades Road Industrial Estate, Oldbury, West Midlands B69 2RA

Tel: +44 (0)121 544 6124 • Fax: +44 (0)121 552 6793

Certificate No. 355b to LPS 1056

LPCB Ref. No.	Spec.	Trade Name	Door Type	Maximum sizes			Fire resistance (min) BS 476 : Part 22	
				Height m	Width m	Area m²	integrity	insulation
355b/01	4	SF Series	rolling shutter	7.00	7.00	49.0	240	0

Northern Doors (UK) Limited

Kingsforth Road, Thurcroft, Rotherham, South Yorks S66 9HU

Tel: +44 (0)1709 545999 • Fax: +44 (0)1709 545341

Certificate No. 003a to LPS 1056

LPCB Ref. No.	Spec.	Trade Name	Door Type	Maximum sizes			Fire resistance (min) BS 476 : Part 22	
				Height m	Width m	Area m²	integrity	insulation
003a/03	4	Fire seal (face-fixed)	rolling shutter	2.30	2.30	5.29	360	0
				7.00	7.00	49.0	240	0
003a/04	4	Escalator	fire shutter	10.00	5.00	50.0	240	0
003a/05	4	Wayfinder	vertically hung rolling shutter	4.50	15.00*	67.50	240	0

* Where this product is used in a firebreak wall or a firebreak floor, insurers may wish to reduce the maximum size.

Syston Rolling Shutters Limited
33 Albert Street, Syston, Leicester LE7 2JB

Tel: +44 (0)116 2608841 • Fax: +44 (0)116 2640846

Certificate No. 039a to LPS 1056

LPCB Ref. No.	Spec.	Trade Name	Door Type	Maximum sizes			Fire resistance (min) BS 476 : Part 22	
				Height m	Width m	Area m²	integrity	insulation
039a/0	4	Spec. 14	face fixed and built-in rolling shutter	7.0	7.0	49.0	240	0

Watson Bros Limited
30-34 Wilson Place, Nerston, East Kilbride G74 4QD

Tel: +44 (0)13552 21232/33144/44476 • Fax: +44 (0)13552 33850

Certificate No. 081a to LPS 1056

LPCB Ref. No.	Spec.	Trade Name	Door Type	Maximum sizes			Fire resistance (min) BS 476 : Part 22	
				Height m	Width m	Area m²	integrity	insulation
081a/01	4	TRD ECOSSE	rolling shutter	2.44	2.13	5.2	240	0

This scheme is currently under preparation. The most important aspect of the LPCB scheme is that the certificated product is a complete door assembly.

Amber Doors Limited

Mason Way, Platts Common Industrial Estate, Hoyland, Barnsley S74 9TG

Tel: +44 (0)1226 351135 • Fax: +44 (0)1226 350176

Certificate No. 001c
Assessed to LPS 1197

General servicing/maintenance	Yes
Maintenance contracts	Yes
24 hour answering service	Yes
24 hour emergency service	Yes

(on site within 4 hours of call in Avon, Bedfordshire, Berkshire, Borders, Cambridgeshire, Cheshire, Cleveland, Clwyd, Cumbria, Derbyshire, Dorset, Dumfries and Galloway, Durham, Essex, Gloucestershire, Gwent, Gwynedd, Hampshire, Hereford and Worcestershire, Hertfordshire, Humberside, Kent, Lancashire, Leicestershire, Lincolnshire, Merseyside, Norfolk, Northamptonshire, Northumberland, Nottinghamshire, Oxfordshire, Powys, Shropshire, Somerset, Staffordshire, Suffolk, Surrey, East Sussex, West Sussex, Tyne & Wear, Warwickshire, West Midlands, Wiltshire, North Yorkshire, South Yorkshire and West Yorkshire).

Ascot Industrial Doors Limited

Britannia Way Industrial Park, Union Road, Bolton BL2 2HE

Tel: +44 (0)990 556644 • Fax: +44 (0)1204 545800

Certificate No. 131b
Assessed to to LPS 1197

General servicing/maintenance	Yes
Maintenance contracts	Yes
24 hour answering service	Yes
24 hour emergency service	Yes

(on site within 4 hours of all in England, Scotland and Wales)

Blount Shutters Limited

Unit B, 734 London Road, West Thurrock, Essex RM16 1NL

Tel: +44 (0)1708 860000 • Fax: +44 (0)1708 861271

Certificate No. 347b
Assessed to LPS 1197

General servicing/maintenance	Yes
Maintenance contracts	Yes
24 hour answering service	Yes
24 hour emergency service	Yes

(on site within 4 hours of call in England and Wales. Isles dependent on ferries)

Bolton Brady Repair and Service Limited

Unit 12, Hunslet Trading Estate, Severn Road, Hunslet, Leeds LS10 1BL

Tel: +44 (0)1132 718633 • Fax: +44 (0)1132 771808

Certificate No. 122a
Assessed to LPS 1197

General servicing/maintenance	Yes
Maintenance contracts	Yes
24 hour answering service	Yes
24 hour emergency service	Yes

(on site within 4 hours of call in Humberside and Yorkshire.)

Bolton Brady Repair and Service Limited
Unit 4, Queensway, Walworth Industrial Estate, Andover SP10 5AZ

Tel: +44 (0)1264 350616 • Fax: +44 (0)1264 336025

Certificate No. 122a
Assessed to LPS 1197

General servicing/maintenance	Yes	
Maintenance Contracts	Yes	
24 hour answering service	Yes	
24 hour emergency service	Yes	(on site within 4 hours of call in Berkshire, Oxfordshire, Hampshire, Surrey and Sussex [west of A24].)

Bolton Brady Repair and Service Limited
18 Power Court, Luton LU1 3JJ

Tel: +44 (0)1582 28607 • Fax: +44 (0)1582 484532

Certificate No. 122a
Assessed to LPS 1197

General servicing/maintenance	Yes	
Maintenance contracts	Yes	
24 hour answering service	Yes	
24 hour emergency service	Yes	(on site within 4 hours of call in Bedfordshire, Buckinghamshire and Hertfordshire)

Bolton Brady Repair and Service Limited
405 Hillington Road, Glasgow G52 4BL

Tel: +44 (0)141 883 2131 • Fax: +44 (0)141 883 4502

Certificate No. 122a
Assessed to LPS 1197

General servicing/maintenance	Yes	
Maintenance contracts	Yes	
24 hour answering service	Yes	
24 hour emergency service	Yes	(on site within 4 hour of call in Borders, Central, Dumfries & Galloway, Fife, Lothian, Strathclyde and South Tayside)

Chiltern Industrial Doors Limited
Unit 11, Commerce Way, Leighton Buzzard, Bedfordshire LU7 8RW

Tel: +44 (0)1525 383537 • Fax: +44 (0)1525 382314

Certificate No. 246a
Assessed to LPS 1197

General servicing/maintenance	Yes	
Maintenance contracts	Yes	
24 hour answering service	Yes	
24 hour emergency service	Yes	(on site within 4 hours of call in: Avon, Bedfordshire, Berkshire, Buckinghamshire, Cambridgeshire, East Sussex, Hampshire, Hertfordshire, Kent, London, Northamptonshire, Oxfordshire, Surrey, West Sussex, Warwickshire and Wiltshire.

(Note : This company also covers the Isle of Wight, depending on the ferry).

2 SECTION 4:
SERVICE AND MAINTENANCE OF DOORS

Dorzone

PO Box 13, Lightwater, Surrey GU18 5PY

Tel: +44 (0)1276 686 655 • Fax: +44 (0)1276 684 411

Certificate No. 424a
Assessed to LPS 1197

General servicing/maintenance	Yes
Maintenance contracts	Yes
24 hour answering service	Yes
24 hour emergency service	Yes (on site within 4 hours of call in area within M25)

Fortress Industries Limited

6 Trench Road, Hyde Park Industrial Estate, Newtownabbey, Co. Antrim, Northern Ireland BT36 8TY

Tel: +44 (0)1232 342655 • Fax: +44 (0)1232 342651

Certificate No. 146b
Assessed to LPS 1197

General servicing/maintenance	Yes
Maintenance contracts	Yes
24 hour answering service	Yes
24 hour emergency service	Yes (on site within 4 hours of call in Northern Ireland.)

Henderson-Bostwick Industrial Doors Limited

Battersea Road, Mersey Industrial Estate, Heaton Mersey, Stockport, Cheshire SK4 3ED

Tel: +44 (0)161 947 4040 • Fax: +44 (0)161 947 4041

Certificate No. 006b
Assessed to LPS 1197

General servicing/maintenance	Yes
Maintenance contracts	Yes
24 hour answering service	Yes
24 hour emergency service	Yes (on site within 4 hours of call in Cheshire, Clwyd, Cumbria, Greater Manchester, Gwynedd, Lancashire, Merseyside, Shropshire and Stafford)

(Note: This company also covers the Isle of Man, depending on the ferry)

Henderson-Bostwick Industrial Doors Limited

26/27 Bates Industrial Estate, Harold Wood, Romford, Essex RM3 0HU

Tel: +44 (0)1708 383700 • Fax: +44 (0)1708 383701

Certificate No. 006b
Assessed to LPS 1197

General servicing/maintenance	Yes
Maintenance contracts	Yes
24 hour answering service	Yes
24 hour emergency service	Yes (on site within 4 hours of call in Bedfordshire, Berkshire, Cambridgeshire, Greater London, Essex, Hertfordshire, Norfolk, Suffolk, Surrey and East Sussex)

Henderson-Bostwick Industrial Doors Limited

Henderson-Lowland, Unit 15, Earn Avenue, Righead Industrial Estate, Bellshill, Strathclyde ML4 3LW

Tel: +44 (0)1698 835400 • Fax: +44 (0)1698 835401

Certificate No. 006b
Assessed to LPS 1197

General servicing/maintenance	Yes	
Maintenance contracts	Yes	
24 hour answering service	Yes	
24 hour emergency service	Yes	(on site within 4 hours of call in Borders, Central, Dumfries and Galloway, Fife, Grampian, Highlands (South of Inverness), Lothian, Strathclyde and Tayside)

Henderson-Bostwick Industrial Doors Limited

Unit 4, Woodham Road, Aycliffe Industrial Estate, Newton Aycliffe, Co Durham DL5 6HT

Tel: +44 (0)1352 303700 • Fax: +44 (0)1352 303701

Certificate No. 006b
Assessed to LPS 1197

General servicing/maintenance	Yes	
Maintenance contracts	Yes	
24 hour answering service	Yes	
24 hour emergency service	Yes	(on site within 4 hours of call in Cleveland, Co. Durham, Northumberland, Tyne and Wear, North and West Yorkshire)

Henderson-Bostwick Industrial Doors Limited

Crow House, Crow Arch Lane Industrial Estate, Ringwood, Hampshire BH24 1PD

Tel: +44 (0)1425 462020 • Fax: +44 (0)1425 462021

Certificate No: 006b
Assessed to LPS 1197

General servicing/maintenance	Yes	
Maintenance contracts	Yes	
24 hour answering service	Yes	
24 hour emergency service	Yes	(on site within 4 hours of call in Avon, Berkshire, Buckinghamshire, Cornwall, Devon, Dorset, Dyfed, West, Mid and South Glamorgan, Gloucestershire, Gwent, Hampshire, Oxfordshire, Powys, Somerset, Surrey, West and East Sussex).

Industrial Door Engineering Limited

Winnington Avenue, Winnington, Northwich, Cheshire CW8 4EQ

Tel: +44 (0)1606 871832 • Fax: +44 (0)1606 871482

Certificate No. 165b
Assessed to LPS 1197

General servicing/maintenance	Yes	
Maintenance contracts	Yes	
24 hour answering service	Yes	
24 hour emergency service	Yes	(on site within 4 hours of call in Cheshire, Clwyd, Derbyshire Greater Manchester, Gwynedd, Lancashire, Merseyside and Yorkshire)

L. B. Securities
Unit 6, Sterling Industrial Estate, Rainham Road South, Dagenham, Essex RM10 8TX

Tel: +44 (0)181 517 6655 • Fax: +44 (0)181 984 0378

Certificate No. 247a
Assessed To LPS 1197

General servicing/maintenance	Yes	
Maintenance contracts	Yes	
24 hour answering service	Yes	
24 hour emergency service	Yes	(on site within 4 hours of call-out in: Bedfordshire, Berkshire, Buckinghamshire, Cambridgeshire, Essex, Hampshire, Hertfordshire, Kent, London, Norfolk, Northamptonshire, Oxfordshire, Suffolk, Surrey, East Sussex, West Sussex)

Lycetts (Burslem) Limited
Glendale Street, Burslem, Stoke on Trent ST6 2EP

Tel: +44 (0)1782 575236 • Fax: +44 (0)1782 577841

Certificate No. 029b
Assessed to to LPS 1197

General servicing/maintenance	Yes	
Maintenance contracts	Yes	
24 hour answering service	Yes	(on site within 24 hours of call.)
24 hour emergency service	No	

Mercian Shutters Limited
Pearsall Drive, Brades Road Industrial Estate, Oldbury, West Midlands B69 2RA

Tel: +44 (0)121 544 6124 • Fax: +44 (0)121 552 6793

Certificate No. 355a
Assessed to LPS 1197

General servicing/maintenance	Yes	
Maintenance contracts	Yes	
24 hour answering service	Yes	
24 hour emergency service	Yes	(on site within 4 hours of call in Hereford and Worcestershire, Shropshire, Staffordshire, Warwickshire and West Midlands).

Northern Doors (UK) Limited
Kingsforth Road, Thurcroft, Rotherham, South Yorks SE66 9HU

Tel: +44 (0)1709 545999 • Fax: +44 (0)1709 545341

Certificate No. 003b
Assessed to LPS 1197

General servicing/maintenance	Yes	
Maintenance contracts	Yes	
24 hour answering service	Yes	(on site within 24 hours of call.)
24 hour emergency service	No	

Stanair Industrial Door Services Limited

2 Henson Way, Telford Industrial Estate, Kettering, Northamptonshire NN6 8PX

Tel: +44 (0)1536 82187 • Fax: +44 (0)1536 411799

Also at: The Old Abattoir, 1a Bradwell Road, New Bradwell, Milton Keynes MK13 0BX
Tel: +44 (0)1908 222070 • Fax: +44 (0)1908 222621

Certificate No. 141a
Assessed To LPS 1197

General servicing/maintenance	Yes	
Maintenance contracts	Yes	
24 hour answering service	Yes	
24 hour emergency service	Yes	(on site within 4 hours of call in Bedfordshire, Buckinghamshire, Cambridgeshire, Leicestershire, Northamptonshire).

Syston Rolling Shutters Limited

33 Albert Street, Syston, Leiceste LE7 2JB

Tel: +44 (0)116 260 8841 • Fax: +44 (0)116 264 0846

Certificate 039b
Assessed to LPS 1197

General servicing/maintenance	Yes	
Maintenance contracts	Yes	
24 hour answering service	Yes	
24 hour emergency service	Yes	(On site within 4 hours of call in Bedfordshire, Buckinghamshire, Cambridgeshire, Derbyshire, Hertfordshire (north of M25), Leicestershire, Lincolnshire, Norfolk (west of A140), Northamptonshire, Nottinghamshire, Oxfordshire, Shropshire, Staffordshire, Suffolk (north-west of A143), Warwickshire and Hereford and Worcestershire (north of A417).

Watson Bros Limited

30-34 Wilson Place, Nerston, East Kilbride G74 4QD

Tel: +44 (0)13552 21232/33144/44476 • Fax: +44 (0)13552 33850

Certificate No. 081b
Assessed to LPS 1197

General servicing/maintenance	Yes	
Maintenance contracts	Yes	
24 hour answering service	Yes	
(0141 641 8187)		
24 hour emergency service	Yes	(On site within 4 hours of call in Cumbria (north of A66), Northumberland and Scotland).

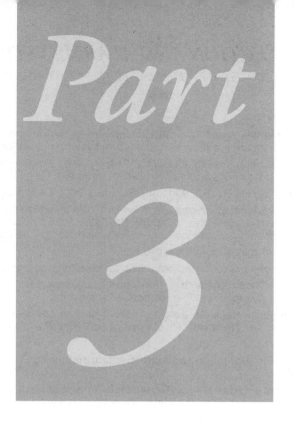

Part 3

FIRE DETECTION AND ALARM SYSTEMS

3 FIRE DETECTION AND ALARM SYSTEMS

SECTION 1: CERTIFICATED FIRE DETECTION AND ALARM SYSTEM FIRMS

INTRODUCTION

The firms listed in this section are Certificated Fire Detection and Alarm System Firms authorised to certificate fire detection and alarm system installations.

These firms have been assessed to *LPS 1014: Issue 3 - Requirements for Certificated Fire Detection and Alarm System Firms,* which requires;

- compliance with *BSEN ISO 9001 or BSEN ISO 9002 Quality Systems,*
- inspection of fire detection and alarm system installations to the requirements of the applicable Installation Rules, primarily *BS 5839: Part 1 - Fire detection and alarm systems for buildings, Code of Practice for system design, installation and servicing.*

Examples of other Installation Rules are;

BS 6266 - Code of practice for fire protection for electronic data processing installations

BS 7273: Part 1 - Code of practice for the operation of fire protection measures, Electrical actuation of gaseous total flooding extinguishing systems.

RLS 1 (LPC Rules) - Rules for automatic fire detection and alarm installations for the protection of property (Schedule for the use of BS 5839: Part 1: 1988.)

Certification of Fire Detection and Alarm System Installations and LPCB Certificates of Conformity

It is a requirement of LPS 1014 that Certificated Firms shall issue an LPCB Certificate of Conformity in respect of each completed installation covered by the scheme.

The issue of an LPCB Certificate of Conformity:

- certifies that the fire detection and alarm system was designed, installed and commissioned in accordance with the Installation Rules applied, and
- ensures that the system is recorded by LPCB as a Certificated Fire Detection and Alarm System.

Additional Offices

Where a firm has more than one office, these offices are also required to meet the appropriate requirements of LPS 1014 to be eligible for certification, and are listed separately.

LPCB Inspection of Fire Detection and Alarm Systems

As part of the ongoing surveillance of a Certificated firm the LPCB selects and inspects completed fire detection and alarm system installations every six months. This is in addition to ISO 9001/9002 surveillance.

PLEASE NOTE: An LPS 1014 Certificate of Conformity will be issued to systems covered by the scheme

ADT Fire and Security
4 Bloomsbury Square, London WC1A 2RL
Tel: +44 (0)171 242 8855 • Fax: +44 (0)171 831 4532
Certificate No. CFA 119

ADT Fire and Security
Security House, The Summit, Hanworth Road, Sunbury on Thames, Middx TW16 5DB
Tel: +44 (0)1932 743333 • Fax: +44 (0)1932 743155
Certificate No. CFA-105

ADT Fire and Security
77/79 Feeder Road, St Philips Marsh, Bristol BS2 0TQ
Tel: +44 (0)117 948 8588 • Fax: +44 (0)117 948 8580
Certificate No. CFA-107

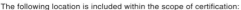
The following location is included within the scope of certification:
Suite C, Phoenix Business Park, Estover Industrial Estate, Estover, Plymouth, Devon PL6 7PY
Tel: +44 (0)1752 738499 • Fax: +44 (0)1752 738985

ADT Fire and Security
ADT House, Mucklow Hill, Halesowen, Birmingham B62 8DA
Tel: +44 (0)121 255 6000 • Fax: +44 (0)121 255 6100
Certificate No: CFA-108

ADT Fire and Security
Quay House, City Park Business Village, Manchester M16 9UN
Tel: +44 (0)161 252 5500 • Fax: +44 (0)161 252 5688
Certificate No. CFA 109

ADT Fire and Security
1 Bowling Green Terrace, Off Jack Lane, Leeds LS11 9SX
Tel: +44 (0)113 244 8441 • Fax: +44 (0)113 242 5933
Certificate No: CFA-110
The following locations are also included within the scope of certification:

Security House, Units 3 & 4 Prior Walk, Saxon Way, Hessle, HU13 9PB
Tel: +44 (0)1482 441310 • Fax: +44 (0)1482 441331
98 Trippet Lane, Sheffield S1 4EL
Tel: +44 (0)114 2797928 • Fax: +44 (0)114 2700818

ADT Fire and Security
Unit 8, Prospect Business Park, Langston Road, Loughton, Essex IG10 3TR
Tel: +44 (0)181 502 5252 • Fax: +44 (0)181 502 2265
Certificate No. CFA112
The following location is included within the scope of certification:
69-75 Thorpe Road, Norwich NR1 1UD
Tel: +44 (0)1603 625138 • Fax: +44 (0)1603 761564

PLEASE NOTE: An LPS 1014 Certificate of Conformity will be issued to systems covered by the scheme

ADT Fire and Security
ADT House, 5th Avenue Business Park, Team Valley Trading Estate, Gateshead NE11 0HF

Tel: +44 (0)191 491 1770 • Fax: +44 (0)191 491 1772

Certificate No. CFA-111

ADT Fire and Security
ADT House, Tannochside Park, Uddingston, Glasgow G71 5PH

Tel: +44 (0)1698 486000 • Fax: +44 (0)1698 486100

Certificate No. CFA 113

BBC Fire Protection Limited
Diamond Road, Norwich, Norfolk NR6 6AW

Tel: +44 (0)1603 486500 • Fax: +44 (0)1603 788957

Certificate No. CFA-126

Caradon Gent Limited
140 Waterside Road, Hamilton Industrial Park, Leicester LE5 1TN

Tel: +44 (0)116 2462000 • Fax: +44 (0)116 2462300

Certificate No. CFA-118

Chubb Electronic Security Limited (North East Region)
5 Canal Place, Armley Road, Leeds LS12 2DU

Tel: +44 (0)113 244 0541 • Fax: +44 (0)113 242 4621

Certificate No: 120

Chubb Electronic Security Limited (London Region)
Security House, Five Ways Business Centre, Aspen Way, Feltham, Middlesex TW13 7AQ

Tel: +44 (0)181 890 8999 • Fax: +44 (0)181 890 8980

Certificate No. CFA-121

Chubb Electronic Security Limited (North West Region)
Unit 7, Park Seventeen, Moss Lane, Whitefield, Manchester M45 8FJ

Tel: +44 (0)161 766 7437 • Fax: +44 (0)161 766 8218

Certificate No. CFA-122

Chubb Electronic Security Limited (Southern Region)
84-86 Jubilee Road, Waterlooville, Hants PO7 7RE

Tel: +44 (0)1705 256588 • Fax: +44 (0)1705 230685

Certificate No. CFA-123

PLEASE NOTE: An LPS 1014 Certificate of Conformity will be issued to systems covered by the scheme

Chubb Electronic Security Limited (Scotland)
186 Garscube Road, Glasgow, Scotland G4 9RQ

Tel: +44 (0)141 332 3230 • Fax: +44 (0)141 332 6128

Certificate No. CFA-124

Dalkia Technical Services Limited
69 Bondway, Vauxhall, London SW8 1SQ

Tel: +44 (0)171 820 3714 • Fax: +44 (0)171 820 3715

Certificate No: CFA-129

Fire Defence plc
Unit 6, Station Road, South Molton, Devon EX36 3LL

Tel: +44 (0)1769 574070 • Fax: +44 (0)1769 574079

Certificate No. CFA-133

F.S.E. Systems Limited
Wilford Industrial Estate, Ruddington Lane, Wilford, Nottingham NG11 7DE

Tel: +44 (0)115 981 2624 • Fax: +44 (0)115 981 6605

Certificate No. CFA-134

Grainger Fire Protection Limited
Thornley House, Carrington Business Park, Carrington, Manchester M31 4XL

Tel: +44 (0)161 777 6700 • Fax: +44 (0)161 777 6638

Certificate No. CFA-125

The following location is included within the scope of certification for servicing:
Unit 1A, Newton Court, Wavertree Technology Park, Wavertree, Liverpool L13 1EJ
Tel: +44 (0)151 220 4068 • Fax: +44 (0)151 259 4365

How Fire Limited
Hillcrest Business Park, Cinderbank, Dudley, West Midlands DY2 9AP

Tel: +44 (0)1384 458993 • Fax: +44 (0)1384 458981

Certificate No. CFA-103

Initial Shorrock Fire
Unit 7, Gregson Industrial Estate, Birmingham Road, Oldbury, West Midlands B69 4EX

Tel: +44 (0)121 552 1105 • Fax: +44 (0)121 544 6642

Certificate No: CFA-130

Initial Shorrock Fire
Unit 1, Grove Industrial Estate, Gloucester Road North, Patchway, Bristol BS34 5BB

Tel: +44 (0)117 9312929 • Fax: +44 (0)117 9236198

Certificate No: CFA-114

PLEASE NOTE: An LPS 1014 Certificate of Conformity will be issued to systems covered by the scheme

Initial Shorrock Fire
Unit 7 Ashton Gate, Ashton Road, Harold Hill, Romford, Essex RM3 8GH

Tel: +44 (0)1708 372020 • Fax: +44 (0)1708 372066

Certificate No: CFA-115

Initial Shorrock Fire
Wilson House, Bramble Farm Business Park, Waterberry Drive, Waterlooville, Hampshire PO7 7XX

Tel: +44 (0)1705 230566 • Fax: +44 (0)1705 230567

Certificate No: CFA-116

Initial Shorrock Fire
Unit 2, Olds Close, Tolpits Lane, Watford, Herts WD1 8RU

Tel: +44 (0)1923 775099 • Fax: +44 (0)1923 770616

Certificate No: CFA-117

Practical Property Protection
92 Cadewell Lane, Shiphay, Torquay, Devon TQ2 7HP

Tel: +44 (0)1803 616623 • Fax: +44 (0)1803 616624

Certificate No. CFA-131

Preussag Fire Protection Limited
Field Way, Greenford, Middlesex UB6 8UZ

Tel: +44 (0)181 832 2000 • Fax: +44 (0)181 832 2200

Certificate No. CFA-104

Protec Fire Detection plc
Protec House, Churchill Way, Nelson, Lancashire BB9 6RT

Tel: +44 (0)1282 717171 • Fax: +44 (0)1282 717273

Certificate No. CFA 132

Raysil Security Systems Limited
Unit 16, Red Lion Business Centre, 218 Red Lion Road, Surbiton, Surrey KT6 7QD

Tel: +44 (0)181 391 0202 • Fax: +44 (0)181 391 4758

Certificate No: CFA-127

Siemens Building Technologies Limited
Cerberus Division, Trinity Court, Woosehill, Wokingham, Berkshire RG41 3AE

Tel: +44 (0)1189 894488 • Fax: +44 (0)1189 775750

Certificate No. CFA-100

PLEASE NOTE: An LPS 1014 Certificate of Conformity will be issued to systems covered by the scheme

Surefire Systems Limited
Marston Court, 98-106 Manor Road, Wallington, Surrey SM6 0DW

Tel: +44 (0)181 773 3770 • Fax: +44 (0)181 669 6652.

Certificate No. CFA-106

TVF (UK) PLC (Also known as TVF Servicetech)
59-69 Queens Road, High Wycombe, Buckinghamshire HP13 6AH

Tel: +44 (0)1494 450641 • Fax: +44 (0)1494 465378

Certificate No: CFA-128

3 SECTION 2:
FIRE DETECTION AND ALARM SYSTEMS

This Section lists Approved and Certificated fire detection and alarm systems. The current requirements for obtaining certification are that all system components shall be listed elsewhere in Part 3, and are shown to be compatible with one another. Assessment of compatibility is performed to the requirements of; Loss Prevention Standard *LPS 1054 - Requirements for the evaluation of component compatibility for fire alarm systems.* Networked systems are evaluated by ensuring that each control panel connected to the network can operate as a standalone unit in the event of failure of the network, and that failure of a control panel or corruption of data by one control panel will not affect other control panels on the network. Where bases are listed with detectors in Section 2, these are the only detector/base combinations that have been shown to be compatible with the control and indicating equipment. Otherwise all detector/base combinations detailed in Section 4 are compatible.

Note

The LPCB approves and certificates products to the requirements of product standards. Where practical, the LPCB, in consultation with the manufacturer, will also verify that the products can meet the recommendations of applicable codes of practice.

The LPCB uses national and international standards for the listing of products. In some instances the requirements of these standards may conflict with the recommendations of local codes of practice. When this situation arises, it is on the requirements of the product standard that the LPCB base approval and certification.

If full compliance with an installation code of practice is necessary, confirmation should be obtained that the listed products also comply with the appropriate codes of practice.

Caradon Gent Limited

140 Waterside Road, Hamilton Industrial Park, Leicester LE5 1TN

Tel: +44 (0)116 2462000 • Fax: +44 (0)116 2462300

Certificate No: 042e to LPS 1054

Certificated System
3260 Fire Alarm System
Comprising the following Gent products:
   13260-01 1 zone control and indicating equipment
   13260-02 2 zone control and indicating equipment
   13260-04 4 zone control and indicating equipment
   13260-08 8 zone control and indicating equipment

Point smoke detectors
   17430-01 ionization smoke detector
   17630-01 ionization smoke detector
   17640-01 photoelectric smoke detector
   72331-25NM ionization smoke detector
   72341-25NM photoelectric smoke detector

Point heat detectors
   17460-01 grade 2 rate of rise heat JŌtector
   17650-01 grade 2 fixed temperature heat detector
   17660-01 grade 1 rate of rise heat detector

Manual call point
   1195OR manual call point

Sounders
   12511-37 alarm sounder
   12511-52 alarm sounder

LPCB

LPCB Ref. No.
042e/S1

Continued

Caradon Gent Limited (continued)

3400 Fire Alarm System   042e/S2
Comprising the following Gent products:
   3404 control and indicating equipment
   3460 mimic panel
   13473-01 analogue ionization smoke sensor
   13471-01 analogue optical smoke/heat sensor
   13472-01 analogue rate of rise heat sensor
   13474 analogue optical beam smoke sensor
   13480-02 manual call point
   13420-02 alarm sounder
   13420-03 alarm sounder

Middle East Sales Enquiries: MK Middle East Marketing Office, Dubai
Tel: +971 4 668 483 • Fax: +971 4 681 585

India Sales Enquiries: MK India, Madras
Tel: +91 44 626 9991 • Fax: +91 44 626 9992

Far East Sales Enquiries: MK Electric (Singapore) Pte. Limited
Tel: +65 271 7266 • Fax: +65 274 9219

Cerberus AG

Nelm AG Factory, Prati San Martino, CH6850 Mendrisio, Switzerland

Tel: +41 1 922 6111 • Fax: +41 1 922 6450
UK Sales: Cerberus Limited
Tel: +44 (0)1734 783703 • Fax: +44 (0)1734 775750

Certificate No. 127d to LPS 1054

LPCB

Approved System   LPCB Ref. No.
S1110 Danger Management System (fire)   127d/S1
Comprising the following Cerberus products
   CS1110 AlgoControl Fire Detection Control Unit

Point smoke detectors
   DO1101A photoelectric smoke detector, collective (DB1101A base)
   F716I ionization smoke detector (Z74A base)
   R716I optical smoke detector (Z74A base)
   F911 ionization smoke detector (Z94C base)

Point heat detectors
   DT1101A garde 1 rate of risse heat detector (DB1101A base)
   DT1102A range 1 rate of rise high temperature heat detector (DB1101A)
   D716 grade 2 rate of rise heat detector (Z74A base)
   D901 grade 1 rate of rise heat detector (Z94C base)
   D921 range 1 rate of rise heat detector (Z94C base)

Manual call point
   DM1101 indoor manual call point (DMZ1191 mounting box)

Flame detectors
   S2406 infra red flame detector
   S2406EX intrinsically safe infra red flame detector.

S1115 AlgoControl Fire Detection System   127d/S2
Comprising the following Cerberus products:
   CS1115 AlgoControl Fire Detection Control Unit

Point smoke detectors:
   DO1101A conventional photoelectric smoke detector, conventional (DB1101A base)
   DO1131A photoelectric smoke detector, addressable (DB1131A base)
   DOT1131A photoelectric smoke detector with thermal enhancement, addressable (DB1131A base)
   F716I ionisation smoke detector (Z74A base)
   R716I optical smoke detector (Z74A base)
   F911 ionisation smoke detector (Z94C base)

Point heat detectors:
   DT 1101A grade 1 rate of rise heat detector (DB1101A base)
   DT 1102A range 1 rate of rise high temperature heat detector, conventional (DB1101A base)
   DT 1131A grade 1 rate of rise heat detector, addressable (DB1131A base)
   D716 grade 2 rate of rise heat detector (Z94C base)
   D901 grade 1 rate of rise heat detector (Z94C base)

Cerberus AG (continued)

D921 range 1 rate of rise heat detector (Z94C base)

Manual call point:
DM1101 manual call point (DMZ1191 mounting box)
DM1131 manual call point (DMZ1191 mounting box)
Beam detectors:
DLO1191 optical beam smoke detector (DLB1191A base)
Flame detectors:
S2406 infra red flame detector
S2406EX intrinsically safe infra red flame detector.
Line module:
DC1131 input module.

Cerberus AG

Alte Landstrasse 411, CH-8708 Männedorf, Switzerland

Tel: +41 1 922 6111 • Fax: +41 1 922 6450
UK Sales: Cerberus Limited: Tel 01734 783703 • Fax: +44 (0)1734 775750.

For listing of LPCB certificated and approved products, please refer to new entry under **'Siemens Building Technologies AG, Cerberus Division'**.

Control Equipment Limited

Hillcrest Business Park, Dudley, West Midlands DY2 9AP

Tel: +44 (0)1384 458651 • Fax: +44 (0)1384 458972

Certificate No: 018b to LPS 1054
Certificated System
Precept fire alarm system
Comprising the following products:
Control Equipment Limited
Precept 2, 4, 8, 16 and 32 zone control and indicating equipment.
Apollo Fire Detectors Limited
55000-100 grade 1 rate of rise heat detector (45681-200 base)
55000-200 ionization smoke detector (45681-200 base)
55000-300 photoelectic smoke detector (45681-200 base)
53531-221 grade 1 rate of rise heat detector (45681-007 base)
53541-151 ionization smoke detector (45681-007 base)
53541-200 photoelectric smoke detector (45681-007 base)
55000-101 grade 2 rate of rise heat detector (45681-200 base)
55000-102 grade 3 rate of rise heat detector (45681-200 base)
55000-103 range 1 rate of rise heat detector (45681-200 base)
55000-104 range 2 rate of rise heat detector (45681-200 base)
55000-210 integrating ionization smoke detector (45681-200 base)
Hochiki Corporation
DFE-60 grade 2 fixed temperature heat detector (YBF-RL/4H5 base)
DFE-90 range 1 fixed temperature heat detector (YBF-RL/4H5 base)
DCC-1EL grade 1 rate of rise heat detector (YBF-RL/4H5 base)
DCC-2EL grade 2 rate of rise heat detector (YBF-RL/4H5 base)
DCC-R1EL range 1 rate of rise heat detector (YBF-RL/4H5 base)
Hochiki Europe
SLK-E photoelectric smoke detector (YBF-RL/4H5 base)
SIH-E ionization smoke detector (YBF-RL/4H5 base)
KAC Alarm Company Limited
KR72 manual call point.
Photain Controls Limited
ISD-P91 ionization smoke detector
PRR-P91 grade 1 rate of rise heat detector
PHD-P90 range 1 fixed temperature heat detector
Signature Industries - Clifford and Snell Alarms.
MCP/R/F manual call point
MCP/R/S manual call point
System Sensor, St. Charles, Illinois, USA.
1151E ionization smoke detector (B401 base)
2151E photoelectric smoke detector (B401 base)

(LPCB)
LPCB Ref. No.
018b/S1

Edwards Systems Technology

6411 Parkland Drive, Sarasota, Florida 34243, USA
Trading through ADT Security Systems for UK and Europe

Tel: +44 (0)171 407 9741 • Fax: +44 (0)171 407 1693

Certificate No 257b to LPS 1054

Certificated system
IRC-3 Fire Alarm System
Comprising the following Edward Systems Technology products:
IRC-3 control and indicating equipment

LPCB Ref No.
257b/S1

Point smoke detectors
1551F analogue ionisation smoke sensor (B501 base)
2551F analogue optical smoke sensor (B501 base)

Line modules
M500MF   monitor module
M500CFS   control module
M500XF   short circuit isolator
M501MF   mini monitor module

Pittway Tecnologica
5551RE grade 1 analogue heat sensor
M500KAC manual call point

For details of approved configurations please refer to ADT documents SF33 and P/N 270239.

Honeywell Incorporated

8500 Bluewater Road, NW Albuquerque, New Mexico, 87121-1958, USA

Tel: +1 505 831 7000 • Fax: +1 505 831 7420
UK Sales Enquiries: Honeywell Control Systems Limited
Tel: +44 (0)1344 826000 • Fax: +44 (0)1344 826240

Approved System
FS-90 analogue addressable fire alarm system
  Comprising the following Honeywell products;
  FS90 control and indicating equipment (to BS 5839: Part 4)

LPC Ref. No.
68/S1

  *For LPC Requirements see Honeywell document reference FS90/LPC/001/0690, Revision 0
  Deltanet Micro Central/Excel Plus System*

Point smoke detectors
  TC807A1036 analogue ionization smoke sensor (14506414-002 base)
  TC806A1037 analogue photoelectric smoke sensor (14506414-002 base)
  TC804E1030 conventional photoelectric smoke detector (14506587-005 base)
  TC805E1013 conventional ionisation detector (14506587-005 base).
  TC807E1003 analogue ionization smoke detector (14506414-007 base)
  TC806E1020 analogue photoelectric smoke detector (14506414-007 base)
  TC806E1012 analogue photoelectric smoke sensor (14506414-007 base)
    Note: Meets BS 5445: Part 7 (EN54 Part 7) with detector 'Low', 'Normal' and 'High' sensitivity setting.
  TC807E1011 analogue ionisation smoke sensor (14506414-007 base)

Point heat detectors
  TC808A1043 grade 1 analogue heat sensor (14506414-002 base)
  TC808E1028 grade 1 analogue heat sensor   (14506414-007 base)
  TC830E1004 grade 1 rate of rise heat detector (14506587-005 base)
  TC808E1002 grade 2 fixed temperature heat detector (14506414-007 Base)

Continued

Honeywell Incorporated (continued)
Line modules
TC809A1059 monitor module
TC810A1056 control module
TC811A1006 short circuit isolator
TC809B1008 mini monitor module
TC809E1019 monitor module
TC810E1008 control module
TC809E1027 mini monitor module
TC811E1007 short circuit isolator

Kidde Fire Protection Limited
12 Nelson Industrial Estate, Cramlington, Northumberland NE23 9BL

Tel: +44 (0)1670 713455 • Fax: +44 (0)1670 735553

Certificate No: 054b to LPS 1054

Certificated Systems:
Series C 2000 analogue addressable fire alarm system[1]
(1) Certificated with Apollo Fire Detectors Series 90 Communications protocol only.

LPCB Ref. No.
054b/S1

Comprising the following products:
Kidde Fire Protection Limited
Series C 2000 1-8 loop control and indicating equipment
Series C 2000 9-16 loop extension panel
Series C 2000 repeater with control functions
PS4 4 amp power supply unit
PP10 10 amp power supply unit

Apollo Fire Detectors Limited
54000-601 analogue temperature sensor
54000-701 analogue ionization smoke sensor
54000-801 analogue photoelectric smoke sensor
54000-901 manual call point
54000-010 short circuit isolator
55000-400 grade 2 analogue temperature monitor
55000-500 analogue ionization smoke monitor
55000-600 analogue photoelectric smoke monitor
55000-910 manual call point

Procyon analogue addressable fire alarm system[1]
(1) Certificated with Apollo Fire Detectors Series 90 Communications protocol only.

054b/S2

Comprising the following products:
Kidde Fire Protection Limited
Procyon 4-8 loop control and indicating equipment
Procyon 9-16 loop extension panel
Procyon repeater with control functions
PS4 4 amp power supply unit
PP10 10 amp power supply unit

Apollo Fire Detectors Limited
54000-601 analogue temperature sensor
54000-701 analogue ionization smoke sensor
54000-801 analogue photoelectric smoke sensor
54000-901 manual call point
54000-010 short circuit isolator
55000-400 grade 2 analogue temperature monitor
55000-500 analogue ionization smoke monitor
55000-600 analogue photoelectric smoke monitor
55000-910 manual call point

Continued

Kidde Fire Protection Limited (continued)

Antares analogue addressable fire alarm system[1]   054b/S3

(1) Certificated with Apollo Fire Detectors Series 90 Communications protocol only.

Comprising the following Products:
Kidde Fire Protection Limited
Antares 2 loop control and indicating equipment

Apollo Fire Detectors Limited
54000-601 analogue temperature sensor
54000-701 analogue ionization smoke sensor
54000-801 analogue photoelectric smoke sensor
54000-901 manual call point
55000-400 grade 2 analogue temperature monitor
55000-500 analogue ionization smoke monitor
55000-600 analogue photoelectric smoke monitor
55000-910 manual call point
54000-010 short circuit isolator

Sirius conventional fire alarm system   054b/S4

Comprising the following products:
Kidde Fire Protection Limited
Sirius 2, 4, 8, 16 and 32 zone control and indicating equipment.

Apollo Fire Detectors Limited
55000-100 grade 1 rate of rise heat detector (45681-200 base)
55000-200 ionization smoke detector (45681-200 base)
55000-300 photoelectic smoke detector (45681-200 base)
53531-221 grade 1 rate of rise heat detector (45681-007 base)
53541-151 ionization smoke detector (45681-007 base)
53541-200 photoelectric smoke detector (45681-007 base)
55000-101 grade 2 rate of rise heat detector (45681-200 base)
55000-102 grade 3 rate of rise heat detector (45681-200 base)
55000-103 range 1 rate of rise heat detector (45681-200 base)
55000-104 range 2 rate of rise heat detector (45681-200 base)
55000-210 integrating ionization smoke detectorr (45681-200 base)

Hochiki Corporation
DFE-60 grade 2 fixed temperature heat detector (YBF-RL/4H5 base)
DFE-90 range 1 fixed temperature heat detector (YBF-RL/4H5 base)
DCC-1EL grade 1 rate of rise heat detector (YBF-RL/4H5 base)
DCC-2EL grade 2 rate of rise heat detector (YBF-RL/4H5 base)
DCC-R1EL range 1 rate of rise heat detector (YBF-RL/4H5 base)

Hochiki Europe
SLK-E photoelectric smoke detector (YBF-RL/4H5 base)
SIH-E ionization smoke detector (YBF-RL/4H5 base)

KAC Alarm Company Limited
KR72 manual call point.

Photain Controls Limited
ISD-P91 ionization smoke detector
PRR-P91 grade 1 rate of rise heat detector
PHD-P90 range 1 fixed temperature heat detector

Signature Industries - Clifford and Snell Alarms.
MCP/R/F manual call point
MCP/R/S manual call point

System Sensor, St. Charles, Illinois, USA.
1151E ionization smoke detector (B401 base)
2151E photoelectric smoke detector (B401 base)

3 SECTION 2:
FIRE DETECTION AND ALARM SYSTEMS

Morley Electronic Fire Systems Limited

Morley House, West Chirton, North Shields, Tyne & Wear NE29 7TY

Tel: +44 (0)191 257 6364 • Fax: +44 (0)191 257 6373

LPCB Ref No.
252b/S1

ZXMi-EN analogue addressable fire alarm system[1]
[1]Certificated with Apollo XP95 communication protocol only
Comprising the following products.
Morley Electronic Fire Systems Limited
  ZXMi-EN 1-5 loop control and indicating equipment[1]
Apollo Fire Detectors Limited
  55000-700 short circuit isolator (45681-211 base)
  55000-400 grade 2 analogue temperature monitor (45681-210 base)
  55000-500 analogue ionisation smoke monitor (45681-210 base)
  55000-600 analogue photoelectric smoke monitor (45681-210 base)
  55000-910 manual call point
C-380.LPC analogue addressable fire alarm system[1].
[1]Certificated with Apollo XP95 communication protocol only
Comprising the following products.
Morley Electronic Fire Systems Limited
  C-380.LPC 1-5 loop control and indicating equipment.
  Note: Morley C-380.LPC is only manufactured as Chubb ControlMaster 380
Apollo Fire Detectors Limited
  55000-700 short circuit isolator (45681-211 base)
  5000-400 grade 2 analogue temperature monitor (45681-210 base)
  5000-500 analogue ionisation smoke monitor (45681-210 base)
  5000-600 analogue photoelectric smoke monitor (45681-210 base)
  5000-910 manual call point

252b/S2

Notifier Limited

Charles Avenue, Burgess Hill, West Sussex RH15 9UF

Tel: +44 (0)1444 230300 • Fax: +44 (0)1444 230888

Certificate No: 154b to LPS 1054
Certificated System:
ID1000 analogue addressable fire alarm system

LPCB Ref. No.
154b/S1

Comprising the following Products:
  Notifier Limited
  ID1001/16 1 loop, 16 zone control and indicating equipment
  ID1002/16 2 loop, 16 zone control and indicating equipment
  ID1002/80 2 loop, 80 zone control and indicating equipment
  ID1004/80 4 loop, 80 zone control and indicating equipment
  CRP/16 16 zone repeater panel
  CRP/80 80 zone repeater panel
  CPX-551 analogue ionization smoke sensor
  SDX-551 analogue photoelectric smoke sensor
  FDX-551R grade 1 analogue heat sensor
  CPX-551E analogue ionisation smoke sensor (B501 base)
  FDX-551RE grade 1 analogue heat sensor (B501 base)
  FDX-551E grade 2 analogue heat sensor (B501 base)
  CP-651E conventional ionisation detector.
  CPX-751E analogue ionisation detector.
  SD-651E conventional photoelectric smoke detector (B401 base)
  SDX-751E analogue photoelectric smoke sensor (B501 base)
    Note: Meets BS 5445: Part 7 (EN54 Part 7) with detector 'Low', 'Normal' and 'High' sensitivity setting.
  SDX-551HRE analogue photoelectric smoke sensor (B501 base)
  MMX-1 monitor module
  CMX-2 control module
  MMX-101 mini monitor module
  ISO-X short circuit isolator
  MMX-1E monitor module
  CMX-2E control module
  MMX-101E mini monitor module
  ISO-XE short circuit isolator

  Signature Industries Limited - Clifford and Snell Alarms.
  MCP/R/F manual call point
  MCP/R/S manual call point

Photain Controls
Rudford Estate, Ford Aerodrome, Arundel, West Sussex BN18 0BE

Tel: +44 (0)1903 721531 • Fax: +44 (0)1903 726795

Certificate No: 066e to LPS1054

Certificated Systems
PCS-HR Conventional Fire Detection System
Comprising the following Photain Controls products.
   PCS1200HR and PCS12HRSPDBX 12 zone control and indicating equipment
   PCS1600HR and PCS16HRSPDBX 16 zone control and indicating equipment
   PCS2400HR and PCS24HRSPDBX 24 zone control and indicating equipment
   PCS3600HR and PCS36HRSPDBX 36 zone control and indicating equipment

LPCB Ref.
066e/S1

Point smoke detectors
   ISD-P91 ionisation smoke detector (FSB-P91 base)

Point heat detectors
   PRR-P91 grade 1 rate of rise heat detector (FSB-P91 base)
   PHD-P90 range 1 fixed temperature heat detector (FSB-P91 base)

Protec Fire Detection plc
Protec House, Churchill Way, Nelson, Lancashire BB9 6RT

Tel: +44 (0)1282 692621 • Fax: +44 (0)1282 602570

Certificate No: 201a to LPS1054

Certificated System

AN95 analogue addressable fire alarm system
Comprising the following products:
Protec Fire Detection Plc.
AN95 control and indicating equipment
AN/AD Mark 4 short circuit isolator

LPCB Ref. No.
201a/S1

Nittan (UK) Limited
NID-58-AS-2LR analogue ionisation smoke sensor
2KC-AS-2LR analogue photoelectric smoke sensor
TCA-AS-2LR analogue heat sensor
   Note: Normal sensitivity only.
NCP-AS-2LU manual call point

Siemens Building Technologies AG, Cerberus Division
Alte Landstrasse 411, CH-8708 Männedorf, Switzerland
Manufacturing at: Siemens Building Technologies AG, Cerberus Division, Volketswil Factory, Industriestrasse 22, CH-8604 Volketswil, Switzerland

Tel: +41 1 922 6111 • Fax: +41 1 922 6450
UK Sales: Cerberus Limited:
Tel: +44 (0)1734 783703 • Fax: +44 (0)1734 775750

LPCB

LPC Ref. No.
18/S2

Approved System:
CZ10 Danger Management Control System (Fire)
   Comprising the following Cerberus products
   CZ10 Danger Management Control Unit (Fire)

Point smoke detectors
   F910 ionization smoke detector
   F906 ionization smoke detector
   F911 ionization smoke detector
   R716I optical smoke detector
   F716I ionization smoke detector
   F930 ionization smoke detector
   R930 photoelectric smoke detector

Continued

Siemens Building Technologies AG, Cerberus Division (continued)

Point heat detectors
D900 grade 1 rate of rise heat detector
D920 range 1 rate of rise heat detector
D901 grade 1 rate of rise heat detector
D921 range 1 rate of rise heat detector

Flame detectors
S610 infrared flame detector
S2406 infrared flame detector
S2406Ex infrared flame detector

Manual call point
ATAN 50 manual call point

Line modules
E90 MI master element
M5M 010 Multimaster

Certificate Number 126i to LPS 1054

Certificated System
S1140 analogue addressable fire alarm system

LPCB Ref.No.
126i/S1

Comprising the following products
CS1140 Control and indicating equipment
For details of approved control and indicating equipment, refer to Cerberus AG certificate 126h.
Using E3M060 zone module 'MS91'
E90MI Master element
F716I Ionisation smoke detector (Z94I and Z94MI bases)
F906 Ionisation smoke detector (Z94I and Z94MI bases)
F910 Ionisation smoke detector (Z94I and Z94MI bases)
F930 Ionisation smoke detector (Z94I and Z94MI bases)
R716I Optical smoke detector (Z94I and Z94MI bases)
R930 Optical smoke detector (Z94I and Z94MI bases)
D900 Grade 1 rate of rise heat detector (Z94I and Z94MI bases)
D920 Range 1 rate of rise heat detector (Z94I and Z94MI bases)
ATAN 50Manual call point

Using E3M080 group line unit
F716I Ionisation smoke detector (Z94, Z94B and Z94D bases)
F906 Ionisation smoke detector (Z94, Z94B and Z94D bases)
F910 Ionisation smoke detector (Z94, Z94B and Z94D bases)
F930 Ionisation smoke detector (Z94, Z94B and Z94D bases)
R716I Optical smoke detector (Z94, Z94B and Z94D bases)
R930 Optical smoke detector (Z94, Z94B and Z94D bases)
R936 Optical smoke detector (Z94, Z94B and Z94D bases)
D900 Grade 1 rate of rise heat detector (Z94, Z94B and Z94D bases)
D920 Range 1 rate of rise heat detector (Z94, Z94B and Z94D bases)
S610 Infra red flame detector (FKS6.1 base)
S2406 Infra red flame detector (Z2406 base)
S2406 Ex Intrinsically safe infra red flame detector (Z2406 base)
ATAN 50 Manual call point
DO1101A Photoelectric smoke detector (DB1101A base)
DT1101A Grade 1 rate of rise heat detector (DB1101A base)
DT1102A Range 1 rate of rise high temperature heat detector (DB†101A base)
DM1101 Manual call point
DLO1191 Optical beam smoke detector (DLB1191A base)

Using E3M070 line module 'interactive'
DO1151A Photoelectric smoke detector (DB1151A base)
DO1152A Photoelectric smoke detector (DB1151A base)
DOT1151A Photoelectric smoke detector with thermal enhancement (DB1151A base)
DOT1152A Photoelectric smoke detector with thermal enhancement (DB1151A base)
DT1152 Rate of rise heat detector (DB1151, DBZ1191 and DBZ1192 bases)

Continued

Siemens Building Technologies AG, Cerberus Division (continued)

DT1152A Rate of rise heat detector (DB1151A base)
DM1151 Manual call point
DC1151Input module
DC1154Output module
DLO1191 Optical beam smoke detector (DLB1191A base)

Using E3M071 line module 'interactive'
DO1151A Photoelectric smoke detector (DB1151A base)
DO1152A Photoelectric smoke detector (DB1151A base)
DOT1151A Photoelectric smoke detector with thermal enhancement (DB1151A base)
DOT1152A Photoelectric smoke detector with thermal enhancement (DB1151A base)
DT1152 Rate of rise heat detector (DB1151, DBZ1191 and DBZ1192 bases)
DT1152A Rate of rise heat detector (DB1151A base)
DM1151 Manual call point
DC1151Input module
DC1154Output module
DC1157Input module
DLO1191 Optical beam smoke detector (DLB1191A base)

Using E3M110 line module
DO1131A Photoelectric smoke detector (DB1131A base)
DT1131A Grade 1 rate of rise heat detector (DB1131A base)
DM1131 Manual call point
DC1131Input module
DC1134Output module

Thorn Security Limited
160 Billet Road, Walthamstow, London E17 5DR

Tel: +44 (0)181 919 4000 • Fax: +44 (0)181 919 4040
Sales enquiries: Tel: +44 (0)1932 743333 • Fax: +44 (0)1932 743155

Certificate No: 143b to LPS 1054

	LPCB Ref. No.
Certificated System	
Minerva conventional fire alarm system	143b/S1

Comprising the following Thorn Security products
Minerva 1, 2 and 4 zone control and indicating equipment

Point smoke detectors
MR301T High Performance Optical (HPO) smoke detector
MF301 ionization smoke detector
MR301 photoelectric smoke detector
MF301D ionization smoke detector
MF301DH ionization smoke detector

Point heat detectors
MD301 grade 1 rate of rise heat detector
MD311 grade 2 fixed temperature heat detector

Excelsior conventional fire alarm system	143b/S2

Comprising the following Thorn Security products
Excelsior 1, 2 and 4 zone control and indicating equipment

Point smoke detectors
MR301T High Performance Optical (HPO) smoke detector
MF301 ionization smoke detector
MR301 photoelectric smoke detector
MF301D ionization smoke detector
MF301DH ionization smoke detector

Continued

Thorn Security Limited (continued)

Point heat detectors
MD301 grade 1 rate of rise heat detector
MD311 grade 2 fixed temperature heat detector

Approved Systems
Minerva analogue addressable fire alarm system
Comprising the following Thorn Security products
Minerva 16 controller
Minerva 16E controller

LPC Ref. No.
10/S11

Minerva 80 controller
Minerva power supply unit
Minerva 16R repeater
Minerva 16ER repeater
Minerva 80R repeater

Point smoke detectors
MF301 ionization smoke detector
MR301 photoelectric smoke detector
MF301D ionization smoke detector
MF301DH ionization smoke detector
MR301T High Performance Optical (HPO) smoke detector
MR401 addressable photoelectric smoke detector
MF401 addressable ionization smoke detector
MR501 analogue photoelectric smoke sensor
MF501 analogue ionization smoke sensor
MR501T High Performance Optical (HPO) analogue smoke sensor

Minerva analogue addressable fire alarm system (continued)

Point heat detectors
MD301 grade 1 rate of rise heat detector
MD311 grade 2 fixed temperature heat detector
MD401 addressable grade 1 rate of rise heat detector
MD501 analogue heat sensor

Flame Detectors
MS502Ex intrinsically safe flame detector

Line Modules
RM520 relay module
SM520 addressable sounder driver module
DM520 conventional detector module
LI520 line isolator module
CM520 contact monitoring module
SB520 sounder booster module

ThornNet System
In conjunction with the following LPCB approved control and indicating equipment.
Minerva 8
Minerva 16E
Minerva 80
When incorporating TLK530 network interface module.

143h/01

Wormald Signalco A/S, Electronics Division

Stanseveien 4, Postboks 52, Kalbakken, 0901 Oslo 9, Norway

Tel: +47 91 76 00 • Fax: +47 91 76 01

Approved System
PBS-16 analogue addressable fire alarm system

LPCB Ref.No
175a/S1

Comprising the following products:
PBS-16 control and indicating equipment (to BS 5839 : Part 4)
US16 expansion unit (to BS 5839 : Part 4)
EL24/3 power supply unit
AX 87IS short circuit isolator
AX 87AD address unit

Nittan (UK) Limited
NID 58 AW analogue ionisation smoke sensor
2KC AW analogue photoelectric smoke sensor
TCA AW analogue heat sensor

Ziton SA (Pty) Limited

Ziton House, 9 Buitenkant Street, Cape Town 8000, South Africa

UK Sales enquiries: Ziton Limited
Tel: +44 (0)1908 281981 • Fax: +44 (0)1908 282554

Certificate No. 092f to LPS 1054

Certificated System
ZP5 Mark 5 analogue addressable fire alarm system.

LPCB Ref No. 092f/S1

Comprising the following Ziton SA (Pty) Limited products
ZP5 Mark 5 control and indicating equipment

Point smoke detectors
ZP710-2 analogue optical smoke sensor (ZP7-SB1 base)
ZP730-2 analogue optical smoke sensor (ZP7-SB1 base)

Point heat detectors
ZP720-2 grade 2 analogue heat sensor (ZP7-SB1 base)

Manual call points
ZP785-2 addressable flush and surface mounted call point.

3 SECTION 3:
CONTROL AND INDICATING EQUIPMENT

INTRODUCTION

Fire alarm control and indicating equipment is evaluated to the requirements of:
BS 5839: Part 4 - Specification for control and indicating equipment; or
BS EN54: Part 2 - control and indicating equipment; and
BS EN54: Part 4 - power supplies.

BS 5839: Part 4 will be withdrawn as a British Standard by April 1999 and applications for certification will no longer be accepted after this date.

Where the control and indicating equipment of modular construction, and the use of printed circuit boards (PCBs) is optional, these are listed with the applicable equipment.

Audit testing

Once listed, control and indicating equipment becomes eligible for audit testing to ensure continued compliance with the applicable product standard.

Note

The LPCB approves and certificates products to the requirements of product standards. Where practical, the LPCB, in consultation with the manufacturer, will also verify that the products can meet the recommendations of applicable codes of practice.

The LPCB uses national and international standards for the listing of products. In some instances the requirements of these standards may conflict with the recommendations of local codes of practice. When this situation arises, it is on the requirements of the product standard that the LPCB base approval and certification.

If full compliance with an installation code of practice is necessary, confirmation should be obtained that the listed products also comply with the appropriate codes of practice.

Caradon Gent Limited

140 Waterside Road, Hamilton Industrial Park, Leicester LE5 1TN

Tel: +44 (0)116 2462000 • Fax: +44 (0)116 2462300

Certificate No: 042d to BS 5839: Part 4

Certificated Products	LPCB Ref. No.
13260-01 1 zone control and indicating equipment	042d/01
13260-02 2 zone control and indicating equipment	042d/02
13260-04 4 zone control and indicating equipment	042d/03
13260-08 8 zone control and indicating equipment	042d/04
3404 control and indicating equipment	042d/05
Incorporating as modular units	
13430-01V3 Local Controller Card (LCC)	
13431-01V3 Loop Processor Card (LPC)	
13432-03V3 Input Output Card (IOC)	
13433-01V3 Random Access Memory (RAM) card	
3460 Mimic Panel	042d/06

Middle East Sales Enquiries: MK Middle East Marketing Office, Dubai
Tel: +971 4 668 483 • Fax: +971 4 681 585

India Sales Enquiries: MK India, Madras
Tel: +91 44 626 9991 • Fax: +91 44 626 9992

Far East Sales Enquiries: MK Electric (Singapore) Pte. Limited
Tel: +65 271 7266 • Fax: +65 274 9219

Cerberus AG

Nelm AG Factory, Prati San Martino, CH6850 Mendrisio, Switzerland

Tel: +41 1 922 6111 • Fax: +41 1 922 6450
UK Sales: Cerberus Limited Tel: +44 (0)1734 783703 • Fax: +44 (0)1734 775750

Certificate No. 127c to pr EN 54: Parts 2 and 4: 1992

Approved Products   LPCB Ref. No.
CS1110 AlgoControl Fire Detection Control Unit   127c/01
Incorporating:
  K3X010 - Master card collective
  K3M010- Line extension card
  K3M020- Line extension card
  B3R080 Parallel indicator card
Approved with the following options from pr EN 54 : 2
  7.8 Output to fire alarm devices
  7.9 Output to fire alarm routing equipment
  7.11 Delays to outputs
  7.13 Alarm counter
  8.9 Output to fault warning routing equipment
  10 Test condition

CS1115 AlgoControl Fire Detection Control Unit   127c/02
Incorporating:
  K3X020 - Master Board 'Analogue Plus'
  K3M010- Line extension card
  K3M020- Line extension card
  Z3S020 - key switch 'nordic'
  Z3S030 - key switch 'KABA'
  B3R080 parallel indicator card
  B3R070 parallel indicator card
Approved with the following options from pr EN 54 : 2
  7.8 Output to fire alarm devices
  7.9 Output to fire alarm routing equipment
  7.11 Delays to outputs
  7.13 Alarm counter
  8.3 Fault signals from points
  8.4 Total loss of the power supply
  8.9 Output to fault warning routing equipment
  10 Test condition

Cerberus AG

Alte Landstrasse 411, CH-8708 Männedorf, Switzerland

Tel: +41 1 922 6111 • Fax: +41 1 922 6450
UK Sales: Cerberus Limited: Tel: +44 (0)1734 783703 • Fax: +44 (0)1734 775750

For listing of LPCB certificated and approved products, please refer to new entry under 'Siemens Building Technologies AG, Cerberus Division'.

Control Equipment Limited

Hillcrest Business Park, Dudley, West Midlands DY2 9AP

Tel: +44 (0)1384 458651 • Fax: +44 (0)1384 458972

Certificate No: 018a to BS 5839: Part 4
Certificated Products   LPCB Ref. No.
Precept 2 zone control and indicating equipment   018a/01
  Incorporating as modular units
    C1423 Motherboard
    C1440 Display Board
Precept 4 zone control and indicating equipment   018a/02
  Incorporating as modular units
    C1423 Motherboard
    C1441 Display Board   *Continued*

Control Equipment Limited (continued)

Precept 8 zone control and indicating equipment	018a/03
Incorporating as modular units	
C1427 Motherboard	
C1462 Display Board	
Precept 16 zone control and indicating equipment	018a/04
Incorporating as modular units	
C1425 Motherboard	
C1426 Display Board	
Precept 32 zone control and indicating equipment	018a/05
Incorporating as modular units	
C1425 Motherboard	
C1426 Display Board	
C1464 Display Board	
Series 600: 2 zone control and indicating equipment	018a/06
Incorporating as modular units	
C1423 Motherboard	
C1440 Display Board	
Series 600: 4 zone control and indicating equipment	018a/07
Incorporating as modular units	
C1423 Motherboard	
C1441 Display Board	
Series 600: 8 zone control and indicating equipment	018a/08
Incorporating as modular units	
C1427 Motherboard	
C1462 Display Board	
Series 600: 16 zone control and indicating equipment	018a/09
Incorporating as modular units	
C1425 Motherboard	
C1426 Display Board	
Series 600: 32 zone control and indicating equipment	018a/10
Incorporating as modular units	
C1425 Motherboard	
C1426 Display Board	
C1464 Display Board	

Edwards Systems Technology

6411 Parkland Drive, Sarasota, Florida 34243, USA

Trading through ADT Security Systems for UK and Europe.

Tel: +44 (0)171 407 9741 • Fax: +44 (0)171 407 1693

Certificate No: 257a to BS 5839 Part 4

Certificated Products

IRC-3 control and indicating equipment
Incorporating as modular units

CM1N Control Unit
Control module motherboard.
ZB8 - 2 zone card.
ZB8 - 5 zone card.
ZB8 - 8 zone card.
ZB0 - 8 relay card
ZAS-1 addressable analogue interface card.
Control module CPU card.
COMM-3 RS-485/RS-232 communication card.
Manual control/indicator PCB.
LCD PCB.

CM1 Control Unit
Control module motherboard.
ZB8 - 2 zone card.
ZB8 - 5 zone card.
ZB8 - 8 zone card.
ZB0 - 8 relay card

LPCB Ref. No
257a/01

Continued

Edwards Systems Technology (continued)

ZAS-1 addressable analogue interface card.
Control module CPU card.
COMM-3 RS-485/RS-232 communication card.
Manual control/indicator PCB.
LCD PCB.
CM2N Control Unit
  Motherboard.
  ZB8 - 2 zone card.
  ZB8 - 5 zone card.
  ZB8 - 8 zone card.
  ZB0 - 8 relay card
  ZAS-1 addressable analogue interface card.
  Control module CPU card.
  COMM-3 R5-485/RS-232 communication card.
  Display.
CM2 Control Unit
  Motherboard.
  ZB8 - 2 zone card.
  ZB8 - 5 zone card.
  ZB8 - 8 zone card.
ZB0 - 8 relay card
  ZAS-1 addressable analogue interface card.
  Control module CPU card.
  COMM-3 R5-485/RS-232 communication card.
SAN Modules
  SDR-32 remote annunciator lamp driver module.
  SDR-32K relay driver module.
  SLU-16R remote annunciator module.
  SLU-16Y remote annunciator module.
  SWU-8 remote annunciator lamp and switch module.
  SAN-CPU annunciator controller card.
IOP-3 RS232 isolator card.
SAN COM R remote control module
PS4A-220 power supply unit.
APS4A-220 power supply unit.
For details of approved configurations please refer to ADT documents SF33 and P/N 270239.

Honeywell Incorporated
8500 Bluewater Road, NW Albuquerque, New Mexico, 87121-1958, USA

Tel: +1 505 831 7000 • Fax: +1 505 831 7420
UK Sales Enquiries: Honeywell Control Systems Limited
Tel: +44 (0)1344 656000 • Fax: +44 (0)1344 656240

Approved Products	LPC Ref. No.
FS90 control and indicating equipment (to BS 5839: Part 4)	68/1
For LPC Requirements see Honeywell document reference FS90/LPC/001/0690, Revision 0	
Deltanet Micro Central/Excel Plus System	68/2

Kidde Fire Protection Limited
Unit 12, Nelson Industrial Estate, Cramlington, Northumberland NE23 9BL

Tel: +44 (0)1670 713455 • Fax: +44 (0)1670 735553

Certificate No: 054c to BS 5839: Part 4
Certificated Products

	LPCB Ref. No.
Series C 2000 1-8 loop control and indicating equipment	054c/01
Incorporating as modular units	
21-000031-000 printer	
21-010002-000 VFD display PCB	
29-000117-000 multiway zonal alarm board 4 way	
29-000117-001 multiway zonal alarm board 8 way	
29-000118-000 zone alarm module card	*Continued*

Kidde Fire Protection Limited (continued)

29-000252-000 twin relay card
29-020325-000 12 way relay card
29-020372-000 power supply PCB
29-020397-000 Hochiki loop card
29-020445-000 interface PCB
29-000456-000 12 way relay card common
29-020460-000 Main processor
29-020489-000 output drive
29-020490-000 output control
29-020498-000 Sabre printer interface
29-020499-000 Apollo loop card
29-020510-000 basic loop controller processor, Hochiki
29-020510-001 basic loop controller processor, Apollo
29-020511-000 loop terminal card
29-020517-000 BLC alarm output card
29-020523-000 ADDS to RS422 interface
29-020526-001 serial data output, TTL version
29-020527-000 Universal Serial Communications (USC) card
29-020530-000 80 Zonal LED/VFD Driver
29-020531-000 opto isolator PCB
DC-40F24T15-DH powersolve dc/dc converter

Series C 2000 9-16 loop extension panel   054c/02
   Incorporating as modular units
29-020397-000 Hochiki loop card
29-020499-000 Apollo loop card
29-020510-000 basic loop controller processor, Hochiki
29-020510-001 basic loop controller processor, Apollo
29-020511-000 loop terminal card
29-020531-000 opto isolator PCB
DC-40F24T15-DH powersolve dc/dc converter

Series C 2000 repeater with control functions   054c/03
   Incorporating as modular units
21-000031-000 printer
21-010002-000 VFD display PCB
29-020372-000 power supply PCB
29-020445-000 interface PCB
29-020498-000 Sabre printer interface
29-020524-001 Addressable repeater board
29-020526-001 serial data output, TTL version
29-020530-000 80 Zonal LED/VFD Driver
DC-40F24T15-DH powersolve dc/dc converter

PS4 4 amp power supply unit   054c/04
PP10 10 amp power supply unit   054c/05
Procyon 4-8 loop control and indicating equipment   054c/06
   Incorporating as modular units
21-020002-000 VFD
21-000035-000 Printer
29-000117-000 multiway zonal alarm board 4-way
29-000117-001 multiway zonal alarm board 8-way
29-000118-000 Zone alarm module card
29-000252-000 Twin relay card
29-020325-000 12 way relay card
29-020372-000 Power supply PCB
29-020456-000 12 way relay card common
29-020460-000 Main processor
29-020489-000 Output processor
29-020490-000 Output control
29-020517-000 BLC alarm output card
29-020523-000 ADDS to RS422 interface
29-020526-001 Serial data output, TTL version
29-020527-000 Universal Serial Communications (USC) card
29-020541-000 Status/zonal display board
29-020542-000 64 zone display board
29-020547-000 Powersolve dc/dc converter   *Continued*

Kidde Fire Protection Limited (continued)

29-020548-000 BLC board, Apollo
29-020549-000 Terminal board, Apollo
29-020550-000 Loop driver PCB, Apollo
29-020551-000 BLC board, Hochiki
29-020552-000 Terminal board, Hochiki
29-020553-000 Loop driver PCB, Hochiki

Procyon 9-16 loop extension panel   054c/07
  Incorporating as modular units
    29-020547-000 Powersolve dc/dc converter
    29-020548-000 BLC board, Apollo
    29-020549-000 Terminal board, Apollo
    29-020550-000 Loop driver PCB, Apollo
    29-020551-000 BLC board, Hochiki
    29-020552-000 Terminal board, Hochiki
    29-020553-000 Loop driver PCB, Hochiki

Procyon repeater with control functions   054c/08
  Incorporating as modular units
    21-000035-000 Printer
    29-020444-000 Power supply PCB
    29-020524-001 Addressable repeater board
    29-020526-001 Serial data output, TTL version
    29-020541-000 Status/zonal display board
    29-020542-000 64 zone display board
    29-020547-000 Powersolve dc/dc converter

Antares 2 loop control and indicating equipment   054c/09
  Incorporating as modular units
    21-000035-000 Printer
    21-010002-000 VFD display PCB
    29-020444-000 Power supply PCB
    29-020550-000 Loop driver PCB, Apollo
    29-020553-000 Loop driver PCB, Hochiki
    29-020554-000 Master processor board
    29-020555-000 Slave processor
    29-020556-000 Apollo personality card
    29-020557-000 Hochiki personality car

Sirius 2, 4, 8, 16 and 32 zone control and indicating equipment   054c/10
  Incorporating as modular units
    29-030615-000 2 zone control board
    29-030619-000 2 zone display board
    29-030615-001 4 zone control board
    29-030619-001 4 zone display board
    29-030616-000 8 zone control board
    29-030620-000 8 zone display board
    29-030617-000 16 zone control board
    29-030621-000 16 zone display board
    29-000117-000 Multi-way zonal alarm board 4 way
    29-000117-001 Multi-way zonal alarm board 8 way
    29-000118-000 Zone alarm module card
    29-000252-000 Twin relay card
    29-000325-000 12 way relay card
    29-000456-000 12 way relay card common

Morley Electronic Fire Systems Limited
Morley House, West Chirton, North Shields, Tyne & Wear NE29 7TY

Tel: +44 (0)191 257 6364 • Fax: +44 (0)191 257 6373

Certificate No. 252a to prEN54 : Parts 2 and 4 : 1995
Approved Products

	LPCB Ref. No.
ZXMi -EN 1-5 loop control and indicating equipment	252a/01
C-380.LPC 1-5 loop control and indicating equipment	252a/02

Note: Morley C-380.LPC is only manufactured as Chubb Controlmaster 380

Approved with the following options from prEN 54 : 2 - 1995
7.8 Output to fire alarm devices
7.11 Delays to outputs
7.12 Co-Incidence detection
8.3 Fault signals from points
9.5 Disablement of addressable points
10 Test condition

Notifier Limited
Charles Avenue, Burgess Hill, West Sussex RH15 9UF

Tel: +44 (0)1444 230300 • Fax: +44 (0)1444 230888

Certificate No: 154a to BS5839: Part 4
Certificated Products

	LPCB Ref. No.
ID1001/16 1 loop, 16 zone control and indicating equipment	154a/01
Incorporating as modular units:	
124-132 power supply unit	
124-083 CP6 processor PCB	
124-080 display and control PCB	
124-090 signalling PCB for connection to repeater	
124-065 loop card	
ID1002/16 2 loop, 16 zone control and indicating equipment	154a/02
Incorporating as modular units:	
124-132 power supply unit	
124-083 CP6 processor PCB	
124-080 display and control PCB	
124-090 signalling PCB for connection to repeater	
124-065 loop card	
ID1002/80 2 loop, 80 zone control and indicating equipment	154a/03
Incorporating as modular units:	
124-132 power supply unit	
124-083 CP6 processor PCB	
124-080 display and control PCB	
124-090 signalling PCB for connection to repeater	
124-065 loop card	
124-105 expansion card	
ID1004/80 4 loop, 80 zone control and indicating equipment	154a/04
Incorporating as modular units:	
124-132 power supply unit	
124-083 CP6 processor PCB	
124-080 display and control PCB	
124-090 signalling PCB for connection to repeater	
124-065 loop card	
124-105 expansion card	
CRP/16 16 zone repeater panel	154a/05
CRP/80 80 zone repeater panel	154a/06

Photain Controls plc

Rudford Estate, Ford Aerodrome, Arundel, West Sussex B18 0BE

Tel: +44 (0)1903 721531 • Fax: +44 (0)1903 726795

Certificate No: 066d to BS 5839: Part 4

PIC 9100 2 loop analogue addressable panel[1]   066d/01
    Incorporating as modular units:
        PCB 4616 24 zone LED board
        PCB 4398 PIC 9000 CPU PCB
        PCB 4526 PIC 9000 Series termination board
        PCB 4321 Analogue loop processor PCB
        PCB 4849 Analogue loop driver PCB
        PCB 4882 PIC9PSUPCB2 power supply

PIC 9200 4 loop analogue addressable panel[1]   066d/02
    Incorporating as modular units:
        PCB 4464 32 zone LED board
        PCB 4398 PIC 9000 CPU PCB
        PCB 4526 PIC 9000 Series termination board
        PCB 4321 Analogue loop processor PCB
        PCB 4849 Analogue loop driver PCB
        PCB 4882 PIC9PSUPCB2 power supply

[1]   Certificated for factory default settings mode as described on Appendix 2 of Installations and Operating Instructions, except
for 'Detector Setup' which requires configuration in conjunction with installation specific requirements.

PCS1200HR and PCS12HRSPDBX 12 zone control and indicating equipment   066d/03
    Incorporating as modular units
        PCS1200CPU Microcontroller PCB
        PCS12ZLEDPCB 4 zone indicator PCB
        PCS12ZTERPCB 4 zone termination PCB
        PCS12SRTPCB Common sounder/Relay outputs
        PCS1216PSPCB Power supply

PCS1600HR and PCS16HRSPDBX 16 zone control and indicating equipment   066d/04
    Incorporating as modular units
        PCS1200CPU Microcontroller PCB
        PCS12ZLEDPCB 4 zone indicator PCB
        PCS12ZTERPCB 4 zone termination PCB
        PCS12SRTPCB Common sounder/Relay outputs
        PCS1216PSPCB Power supply

PCS2400HR and PCS24HRSPDBX 24 zone control and indicating equipment   066d/05
    Incorporating as modular units
        PCS1200CPU   Microcontroller PCB
        PCS12ZLEDPCB 4 zone indicator PCB
        PCS12ZTERPCB 4 zone termination PCB
        PCS12SRTPCB Common sounder/Relay outputs
        PCS236PSPCB Power supply

PCS3600HR and PCS36HRSPDBX 36 zone control and indicating equipment   066d/06
    Incorporating as modular units
        PCS1200CPU Microcontroller PCB
        PCS12ZLEDPCB 4 zone indicator PCB
        PCS12ZTERPCB 4 zone termination PCB
        PCS12SRTPCB Common sounder/Relay outputs
        PCS236PSPCB Power supply

Protec Fire Detection plc

Protec House, Churchill Way, Nelson, Lancashire BB9 6RT

Tel: +44 (0)1282 692621 • Fax: +44 (0)1282 602570

Certificate No: 201b to BS5839: Part 4
Certificated products
AN95 control and indicating equipment
  Incorporating as modular units:
    AN90 Processor Board, issue B
    LPC Charger
    AD/AN95 Display Board, issue A
    2-Loop Transmission I/F MK.2, issue A

LPCB Ref. No.
201b/01

Siemens Building Technologies AG, Cerberus Division

Alte Landstrasse 411, CH-8708 Männedorf, Switzerland.

Manufacturing at: Siemens Building Technologies AG, Cerberus Division, Volketswil Factory, Industriestrasse 22, CH-8604 Volketswil, Switzerland

Tel: +41 1 922 6111 • Fax: +41 1 922 6450

UK Sales: Cerberus Limited:

Tel: +44 (0)1734 783703 • Fax: +44 (0)1734 775750

LPCB Ref.
126h/01

Approved Product
CS1140 control and indicating equipment
Comprising the following products:
CC1140 Control and indicating equipment
CT11 Operators console
CI1140 Integrated central unit
Incorporating as modular units:

E3X100	Master module
E3X101	Master module
E3C010	Battery charging module
E3C011	Battery charging module
B2F020	Converter
E3M070	Line module 'Interactive'
E3M071	Line module 'Interactive'
B3Q460	Control console
B3Q320	Fire department control panel (Swiss)
E3G050	Contacts control unit
E3G070	Universal control unit
E3I020	RS232 unit
E3L020	I/O control unit
E3M080	Group line unit
K3R071	Synoptic - Algo pilot
K3R070	Synoptic - Algo pilot
E3M060	MS9i module
E3G091	Remote TX i/face
E3M110	Analogue plus module
B3R050	Parallel indicator module
B3R051	Parallel indicator module
E3G080	Extinguishing module
B3Q440	Extinguishing module
E3G060	Control monitored module
E3H020	C-bus gateway module

Notes:
(1) The following options with requirements are approved with this equipment:
    7.8   Outputs to fire alarm devices
    7.9   Output to fire alarm routing equipment
    7.10 Output to Fire Protection Equipment
    7.11 Delays to outputs
    7.12 Coincidence detection
    7.13 Alarm counter
    8.3   Fault signals from points
    8.4   Complete failure of power supply
    9.5   Disablement of addressable points
    10   Test condition
    11   Standardized input/output interface

(2) Compliance with clause 12.6.2 of prEN54-2 : 1992 requires LCD screen saving feature to be disabled for the alarm condition.

Thorn Security Limited

160 Billet Road, Walthamstow, London E17 5DR

Tel: +44 (0)181 919 4000 • Fax: +44 (0)181 919 4040

Certificate No: 143c to BS 5839: Part 4
Certificated Products

	LPC Ref.No.
Minerva conventional 1, 2 and 4 zone control and indicating equipment	143c/01
Excelsior conventional 1, 2 and 4 zone control and indicating equipment	143c/02
Minerva 8 addressable/analogue addressable controller (to BS5839 part 4)	143c/03

Incorporating as modular units
MP main processor board
MPIM multipurpose interface board
16ZD 16 way zonal display board
ACPM AC/DC power module.
TLK530 network interface module (ThornNet).
LPIM loop powered insertion module.

Approved Products

	LPC Ref No.
Minerva 16E addressable/ analogue addressable controller (to BS 5839: Part 4)	10/70

Incorporating as modular units
MP main processor board
MPIM multipurpose interface board
16ZD 16 way zonal display board
ACPM AC/DC power module.
TLK530 network interface module (ThornNet).
LPIM loop powered insertion module.

Minerva 80 addressable/ analogue addressable controller (to BS 5839: Part 4)	10/71

Incorporating as modular units
MP main processor board
MPIM multipurpose interface module
80ZD 80-way zonal display board
TLK530 network interface module
MB motherboard
MB-LP motherboard for loop powered systems
AXLM addressable loop expansion module
ALXM-LP expansion module for loop powered sounder.
DCPM DC/DC converter module
DCPM -LP DC power module for loop powering

Minerva power supply unit	10/98

Incorporating ACPM AC/DC power module board

Minerva 16R repeater	10/72

Incorporating as modular units
MPIM multipurpose interface module
16ZD 16-way zonal display board
ACPM AC/DC power module
DCPM DC/DC converter module

Minerva 16ER repeater	10/99

Incorporating as modular units
MPIM multipurpose interface module
16ZD 16-way zonal display board
ACPM AC/DC power module board
DCPM DC/DC converter module

Minerva 80R repeater	10/100

Incorporating as modular units
MPIM multipurpose interface module
80ZD 80-way zonal display board
ACPM AC/DC power module board
DCPM DC/DC converter module

Wormald Ansul (UK) Limited
Wormald Park, Grimshaw Lane, Newton Heath, Manchester M40 2WL

Tel: +44 (0)161 205 2321 • Fax: +44 (0)161 455 4459

Certificate No. 024a to prEN54 Parts 2 and 4: 1995

Approved Products   LPCB Ref. No.
NT 100 8 and 16 zone control and indicating equipment   024a/01

For details of optional functions with requirements see Wormald document 4059/1b/51 dated 5/1/96

Wormald Signalco A/S, Electronics Division
Stanseveien 13, PO Box 52, Kalbakken, 0901 Oslo 9, Norway

Tel: +47 22 91 76 00 • Fax: +47 22 91 76 01

Approved Products   LPCB Ref.No.
PBS-16 control and indicating equipment (to BS 5839 : Part 4)   175b/01
Incorporating as modular units:
  PBS16TE2 motherboard
  PS16 front card
  PBS16A loop card
  COM-16 communication card
  RS-32 zonal indicator board
US16 expansion unit (to BS 5839 : Part 4)   175b/02
Incorporating as modular units:
  PBS16TE2 motherboard
  PBS16SA loop card
  COM-16 communication card

EL24/3 power supply unit   175b/03

Ziton SA (Pty) Limited
Ziton House, 9 Buitenkant Street, Cape Town 8000, South Africa

UK Sales enquiries: Ziton Limited
Tel: +44 (0)1908 281981 • Fax: +44 (0)1908 282554

Certificate No. 092e to BS 5839 : Part 4
Certificated Products

                                                                         LPCB Ref No.
ZP5 Mark 5 2, 4 and 8 loop control and indicating equipment   092e/01
  Incorporating as modular units:
    ZP-PS230V4A-3 power supply unit
    ZP-MCB5 main control board
    ZP-LD8-4 line driver board
    ZP-DU81L-2 LCD
    ZP-DU81P-2 plasma display board
    ZP-PR1 printer kit
    ZP-RL8-1 8 way programmable relay board

INTRODUCTION

This section list all types of approved and certificated fire detectors, covering all principles of detection. The following standards are used for certification schemes.

BS 5445 : Part 5/EN54: Part 5	*Heat sensitive detectors- point detectors containing a static element.*
BS 5445 : Part 7/EN54: Part 7	*Specification for point type smoke detectors using scattered light, transmitted light or ionisation.*
BS 5445 : Part 8/EN54: Part 8	*Specification for high temperature heat detectors.*
BS 5839: Part 5	*Specification for optical beam smoke detectors.*

Applications are currently accepted to the above standards, but with the impending publication of:

prEN 54 Part 5	*Heat detectors - point detectors*
prEN54 Part 7	*Smoke detectors - point detectors using scattered light, transmitted light or ionisation.*

Applications will also be accepted to these standards.

Approval schemes are run for the following products; testing requirements are available from the LPCB.

Flame detectors

Aspirating smoke detectors

Multi criteria detectors

Approval can be obtained for other detection principles not detailed above. In such circumstances , the LPCB develops test criteria to evaluate products, which in time will be used for the development of standards.

The bases with which a detector has been evaluated and shown to meet the relevant standard, are listed with each detector. Bases are not approved/ certificated in their own right.

Audit testing

Once listed, a detector is eligible for audit testing to ensure continued compliance with the applicable product standard.

Note

The LPCB approves and certificates products to the requirements of product standards.

Where practical, the LPCB, in consultation with the manufacturer, will also verify that the products can meet the recommendations of applicable codes of practice.

The LPCB uses national and international standards for the listing of products. In some instances the requirements of these standards may conflict with the recommendations of local codes of practice. When this situation arises, it is on the requirements of the product standard that the LPCB base approval and certification.

If full compliance with an installation code of practice is necessary, confirmation should be obtained that the listed products also comply with the appropriate codes of practice.

3 SECTION 4:
DETECTORS

AirSense Technology Limited
1 Oak House, Knowl Piece, Wilbury Way, Hitchin, Hertfordshire, SG4 OTY

Tel: +44 (0)1462 440 666 • Fax: +44 (0)1462 440 888

LPCB

Certificate No. 404a to GEI 1-048: 30-01-97

Approved Products	**LPCB Ref. No.**
Stratos HSSD Aspirating smoke detector	404a/01

Approved Configuration:
Master Detector and 3 Slave Detectors

The approval is conditional on;
1. A per hole sensitivity of better than 5.00%/m verified during commissioning testing.
2. Power for the aspirating detector shall be supplied by a power supply complying with EN54: Part 4.
3. The use of pipeCAD in designing the sampling pipework system.
4. Alarm latch enable is set in accordance with the manufacturer's specification (function 27).
5. The pipework system shall be produced in High Temp. ABS PA-777D pipe.
6. Interconnection of the Master detector and Slave detectors shall be in accordance with the manufacturer's specification.

Apollo Fire Detectors Limited
36 Brookside Road, Havant, Hampshire PO9 1JR

Tel: +44 (0)1705 492412 • Fax: +44 (0)1705 492754

LPCB

Certificate No: 010a to BS 5445: Part 5
010b to BS 5445: Part 7
010c to BS 5445: Part 8

Certificated Products	**LPCB Ref. No.**
Point Smoke Detectors	
53541-151 ionisation smoke detector (45681-007 base)	010b/01
53541-152 integrating ionisation smoke detector (45681-007 base)	010b/02
54000-701 analogue ionisation smoke monitor (45681-087 base)	010b/03
54000-801 analogue photoelectric smoke monitor (45681-087 base)	010b/04
55000-200 ionisation smoke detector (45681-200 and 45681-201 bases)	010b/05
55000-300 photoelectric smoke detector (45681-200 and 45681-201 bases)	010b/06
55000-210 integrating ionisation smoke detector (45681-200 and 45681-201 bases)	010b/07
55000-500 analogue ionisation smoke monitor (45681-210 base)	010b/08
Note: Certificated with Series 90 and XP95 communication protocols	
55000-600 analogue photoelectric smoke monitor (45681-210 base)	010b/09
Note: Certificated with Series 90 and XP95 communication protocols	
55000-212 intrinsically safe ionization smoke detector (45681-207 Base)[1]	010b/14
55000-213 intrinsically safe integrating ionization smoke detector (45681-207 Base)[1]	010b/15
55000-520 analogue ionisation smoke monitor (45681-210 base)	010b/10
Note: Certificated with Series 90 and XP95 communication protocols	
55000-620 analogue photoelectric smoke monitor (45681-210 base)	010b/11
Note: Certificated with Series 90 and XP95 communication protocols.	
55000-540 analogue intrinsically safe ionisation smoke monitor (45681-215 base)[1]	010b/12
Note: Certificated with Series 90 and XP95 communication protocols.	
55000-640 analogue intrinsically safe photoelectric smoke monitor (45681-215 base)[1]	010b/13
Note: Certificated with Series 90 and XP95 communication protocols.	
Point Heat Detectors	
55000-110 intrinsically safe grade 1 Rate of rise heat detector (60°C) (45681-207 Base)[1]	010a/11
55000-111 intrinsically safe grade 2 Rate of rise heat detector (65°C) (45681-207 Base)[1]	010a/12
55000-112 intrinsically safe grade 3 Rate of rise heat detector (75°C) (45681-207 Base)[1]	010a/13
55000-420 grade 2 analogue temperature monitor (45681-210 base)	010a/09
Note: Certificated with Series 90 and XP95 communication protocols	
55000-440 analogue intrinsically safe grade 2 temperature monitor (45681-215 base)[1]	010a/10
Note: Certificated with Series 90 and XP95 communication protocols.	

Continued

Apollo Fire Detectors Limited (continued)

Point High Temperature Heat Detectors

55000-113	intrinsically safe range 1 Rate of rise heat detector (80°C) (45681-207 Base)[1]	010c/05
55000-114	intrinsically safe range 2 Rate of rise heat detector (100°C) (45681-207 Base)[1]	010c/06

[1] The LPCB certification of these devices does not include the electrical parameters and marking concerning INTRINSIC SAFETY (I.S.) - users should confirm with local regulatory bodies and by consultation with the manufacturer that the devices are separately and correctly IS certificated for their application.

Approved Products	**LPC Ref. No.**
Point Smoke Detectors	
53551-201 photoelectric smoke detector (45681-007 base)	13/13
53551-204 photoelectric smoke detector (45681-007 base)	13/14
Bases	
45681-007 Series 20/30 mounting base	
45681-087 Series 90 mounting base	
45681-200 Series 60 mounting base	
45681-201 Series 60 diode mounting base	
45681-210 XP95 addressable mounting base	
45681-215 intrinsically safe base [1]	
45681-207 base	

Caradon Gent Limited

140 Waterside Road, Hamilton Industrial Park, Leicester LE5 1TN

Tel: +44 (0)116 2462000 • Fax: +44 (0)116 2462300

Certificate No: 042a to BS 5445: Part 5
042b to BS 5445: Part 7
042f to BS 5445: Parts 5 and 7
042g to BS 5445: Parts 5 and 8
042h to BS 5839: Part 5

Certificated Products	**LPCB Ref. No.**
Point Smoke Detectors	
17430-01 ionisation smoke detector (17400-01 base)	042b/01
17630-01 ionisation smoke detector (17400-01, 17600-01 and 17601-01 bases)	042b/02
17640-01 photoelectric smoke detector (17400-01, 17600-01 and 17601-01 bases)	042b/03
72331-25NM ionisation smoke detector (17400-01, 17600-01 and 17601-01 bases)	042b/04
Note: Meets BS 5445: Part 7 (EN54: Part 7) when labelled as per clause 3.2 of the standard.	
72341-25NM photoelectric smoke detector (17400-01, 17600-01 and 17601-01 bases)	042b/05
Note: Meets BS 5445: Part 7 (EN54: Part 7) when labelled as per clause 3.2 of the standard.	
13473-01[1] analogue ionisation smoke sensor (13470-02 and 13470-03 bases)	042b/06
Note: Meets BS 5445: Part 7 when configured to State (0)	
34730 analogue ionisation smoke sensor (34700 and 34761 bases)	042b/07
Note: Meets the requirements of BS 5445 : Part 7 when configured to state (0)	
32730 analogue ionisation smoke sensor (32700 and 19279-01 base)	042b/08
Note: Meets the requirements of BS5445: part 7 when configured to state (0)	
78230-01NM analogue ionisation smoke sensor (78200-01NM and 78261-01NM bases)	042b/09
Note: Meets the requirements of BS5445: part 7 when configured to state (0)	
32715 analogue optical sensor (32700 base)	042b/10
Notes: Meets the requirements of BS5445 Part 7 in state (0).	
78215-01NM anologue optical sensor (78200-01NM and 78261-01 NM bases)	
Note: Meets BS 5445: Part 7 when configured to state (0)	
Point heat detectors	
17460-01 grade 2 rate of rise heat detector (17400-01 base)	042a/01
17650-01 grade 1 fixed temperature heat detector (17400-01, 17600-01 and 17601-01 bases)	042a/02
17660-01 grade 2 rate of rise heat detector (17400-01, 17600-01 and 17601-01 bases)	042a/03
13472-01[1] analogue rate of rise heat sensor (13470-02 and 13470-03 bases)	042g/01
Note: Meets BS 5445: Part 5, grade 2 when configured to State (0), and BS 5445: Part 8, range 1 when configured to State (5)	
34720 analogue rate of rise heat sensor (34700 and 34761 bases)	042a/04
Note: Meets the requirements of BS 5445 : Part 5 grade 1 when configured to state (1) and BS 5445 : Part 5 grade 2 when configured to state (0)	
32720 analogue rate of rise heat sensor (32700 and 19279-01 bases)	042a/05
Note: Meets the requirements of BS5445:Part 5 grade 1 when configured to state (1) and BS5445:Part 5 grade 2 when configured to state (O).	

Continued

Caradon Gent Limited (continued)

78220-01NM analogue rate of rise heat sensor (78200-01NM and 78261-01NM bases)   042a/06
    Note: Meets the requirements of BS5445:Part 5 grade 1 when configured to state (1) and
    BS5445:Part 5 grade 2 when configured to state (O).
Point smoke/heat detectors
13471-01[1] analogue optical smoke/heat sensor (13470-02 and 13470-03 bases)   042f/01
    Note: Meets BS 5445: Part 5, grade 2, and BS 5445: Part 7 when configured to State (0)
34710 analogue optical smoke/heat sensor (34700 and 34710 base)   042f/02
    Note: Meets BS 5445: Part 7 and BS 5445: Part 5 grade 2 in states 0, 8 and 13 and BS 5445:
    Part 7 and BS 5445: Part 5 grade 1 in state 12.
Optical heat sounders
34770 optical heat sounder (34700 and 34710 bases)
    Note: Meets BS 5445: Part 7: 1984/EN54: Part 7: 1982 and BS 5445: Part 5: 1977/EN54: Part 5: 1976 and grade 2
    in states 0, 8 and 13 and EFSG/F/95/007
32775 optical sensor sounder (32700base)   042b/12
    Notes: Meets the requirements of BS5445 Part 7 in state (0).
    Sounder meets LPCB requirements.
78280-01NM optical sensor sounder (78200-01NM and 78261-01NM bases)
    Note: Meets BS 5445: Part 7 when configured to state (0)
    Sounder meets LPCB requirements
Optical Beam Smoke Detectors
13474[1] analogue optical beam smoke sensor (comprises 13474-01 Emitter and 13474-02 Receiver)
(13470-02 and 13470-03 bases)-
    Note: Meets BS 5839: Part 5 when the Emitter is configured to State (0) and the Receiver is   042h/01
    configured to State (2).
    Note: Certificated for use with the 13493-15 angle mounting bracket.

Bases
17400-01 standard base
17600-01 standard base
17601-01 diode base
13470-02[1] 2-way addressable base incorporating short circuit isolation
13470-03[1] 3-way addressable base incorporating short circuit isolation
34700 standard base
34761 flush mounting base
32700 standard base
19279-01 flush mounting base
78200-01NM standard base
78261-01NM flush mounting base

(1) Models manufactured with white mouldings are also Certificated, which have the same model numbers with the suffix 'WH'
    appended.

Middle East Sales Enquiries: MK Middle East Marketing Office, Dubai
Tel: +971 4 668 483 • Fax: +971 4 681 585

India Sales Enquiries: MK India, Madras
Tel: +91 44 626 9991 • Fax: +91 44 626 9992

Far East Sales Enquiries: MK Electric (Singapore) Pte. Limited
Tel: +65 271 7266 • Fax: +65 274 9219

Cerberus AG
Alte Landstrasse 411, CH-8708 Männedorf, Switzerland.

Tel: +41 1 922 6111 • Fax: +41 1 922 6450
UK Sales: Cerberus Limited: Tel: +44 (0)1734 783703 • Fax: +44 (0)1734 775750

For listing of LPCB certificated and approved products, please refer to new entry under **'Siemens Building Technologies AG,
Cerberus Division'.**

Edwards System Technology

6411 Parkland Drive, Sarasota, Florida 34243, USA

Trading through ADT Security Systems for UK and Europe
Tel: +44 (0)171 407 9741 • Fax: +44 (0)171 407 1693

Certificate No. 118b to BS 5445: Part 7
Certificated Products

	LPCB Ref. No.
Point smoke detectors	
1551F analogue ionisation smoke monitor (B501 base)	118b/03
2551F analogue photoelectric smoke monitor (B501 base)	118b/04

GLT Exports Limited

Detection House, 72-78 Morfa Road, Swansea SA1 2EN

Tel: +44 (0)1792 455175 • Fax: +44 (0)1792 455176

Certificate No: 330a to BS5445: Part 7
330b to BS5445: Part 5

Certficated Products	LPCB Ref. No.
Point Smoke Detectors	
ZT600 conventional ionisation smoke detector (ZT110 base)	330a/01
ZTA600 analogue ionisation smoke detector (ZT110 base)	330a/02
ZT601 conventional optical detector (ZT110 base)	330a/03
V301 conventional ionisation smoke detector (ZT110 base)	330a/01
V302 conventional optical smoke detector (ZT110 base)	330a/03
IASCI conventional ionisation smoke detector (ZT110 base)	330a/01
IASI analogue ionisation smoke detector (ZT110 base)	330a/02
IASCO conventional optical detector (ZT110 base)	330a/03
Point Heat Detectors	
ZT602 conventional grade 2 rate of rise heat detector (ZT110 base)	330b/01
ZTA602 analogue grade 2 heat sensor (ZT110 base)	330b/02
IASCH conventional grade 2 rate of rise heat detector (ZT110 base)	330b/01
IASH analogue grade 2 heat detector (ZT110 base)	330b/02
Base ZT110	

Hochiki Corporation

141-1 Maehara, Ejiri, Kakuda-shi, Miyagi 981-15, Japan

Sales Office: Tel: +81 3 3444 4111 • Fax: +81 3 3444 4167
All Enquiries to: Hochiki Europe (UK) Limited
Tel: +44 (0)1634 260131 • Fax: +44 (0)1634 260132

Certificate No: 117a to BS 5445: Part 5
117b to BS 5445: Part 7
117c to BS 5445: Part 8
117d to BS 5839: Part 5

Certificated Products	LPCB Ref No:
Point smoke detectors	
SLK-E photoelectric smoke detector (YBF-RL/4H5, YBF-RL/4[3], YBC-RL/4H5, AMU-B2, AMU-MB, YBF-RL/3J, YCA-RL/3H2[2,3], YCA-RL/5H2[2,3] YBK-RL/4H1A[3] and YBK-RL/4H2[3] base)	117b/01
SIH-E ioni¨âtion smoke detector (YBF-RL/4H5, YBF-RL/4[3], AMU-B2, AMU-MB, YBF-RL/3J, YCA-RL/3H2[2,3] YCA-RL/5H2[2,3], YBK-RL/4H1A[3] and YBK-RL/4H2[3])	117b/02
ALB-E analogue photoelectric smoke sensor (YBF-RL/2NBD base)	117b/03
ALE-E analogue photoelectric smoke sensor (YBJ-RL/2NA base)	117b/04
AIC-E analogue ionisation smoke sensor (YBJ-RL/2NA base)	117b/05
(Meets the requirements of BS5445: Part 7 at one sensitivity setting only and in low power mode)	*Continued*

Hochiki Corporation (continued)

ALG-E analogue photoelectric smoke sensor (YBN-RL/2NA and YBN-R/3[3] bases)	117b/06
AIE-E analogue ionisation smoke detector (YBN-R/2NA and YBN-R/3[3] bases)	117b/07
SLR-E conventional optical smoke detector (YBN-R/4)	117b/08
SIJ-E conventional ionisation smoke detector (YBN-R/4)	117b/09
SLR-E (J) conventional photoelectric smoke detector (YBO-R/4C and YBN-R/4C bases)	117b/10

Point heat detectors

DFE-60 grade 2 fixed temperature heat detector (YBF-RL/4H5, YBC-RL/4H5, AMU-B2, AMU-MB, YBF,RL/3J, YCA-RL/3H2[2, 3], YCA-RL/5H2[2, 3], YBK-RL/4H1A[3] and YBK-RL/4H2[3] bases)	117a/02
DCC-1EL grade 1 rate of rise heat detector (YBC-R/3, YBF-RL/4H5, YBF-RL/4[3], AMU-B2, AMU-MB, YBF-RL/3J, YCA-RL/3H2[2, 3], YCA-RL/5H2[2, 3], YBK-RL/4H1A[3] and YBK-RL/4H2[3] bases)	117a/03
DCC-2EL grade 2 rate of rise heat detector (same bases as for DCC-1EL)	117a/04
ATD-E grade 1 analogue rate of rise heat sensor (YBJ-RL/2NA bases)	117a/05
(also meets the requirements of BS 5445: Part 5 in low power mode)	
ATG-E fixed temperature grade 1 heat detector (YBN-R/2NA and YBN-R/3[3] bases)	117a/06
DFE-90 range 1 fixed temperature heat detector (YBF-RL/4H5, YBC-RL/4H5, AMU-B2, AMU-MB, YBF-RL/3J, YCA-RL/3H2[2, 3], YCA-RL/5H2[2, 3], YBK-RL/4H1A[3] and YBK-RL/4H2[3] bases)	117c/01
DCC-R1EL range 1 rate of rise heat detector (same bases as for DCC-1EL)	117c/02
DCD-1E grade 1 conventional rate of rise heat detector (YBN-R/4 base)	117a/07
DCD-2E grade 2 conventional rate of rise heat detector (YBN-R/4 base)	117a/08
DFJ-60E grade 2 conventional fixed temperature heat detector (YBN-R/4 base)	117a/09
DCD-R1E range 1 conventional rate of rise high temperature heat detector (YBN-R/4 base)	117c/03
DFJ-90E range 1 conventional fixed temperature heat detector (YBN-R/4 base)	117c/04
ATG-E (NP) grade 1 fixed temperature heat detector(YBN-R/2NA and YBN-R/3 bases)	117a/10

Beam Detectors

SPB-E optical beam smoke detector	117d/01
Note: Certificated at a sensitivity setting of 25% and when installed as option 2 as specified in DS0203 issue 1.2.	
SPB-ET optical beam smoke detector	117d/02
Note: Certificated at a sensitivity setting of 25% and with SPB-ET2WI interface kit.	

Approved Products	**LPC Ref No:**
SPA-EA[(1)] optical beam smoke detector	36/53

[1] SPA-EA beam detectors shall be installed in accordance with Hochiki document "Hochiki Photoelectric Beam Detector Model SPA-EA Instruction Manual".

Bases
YBF-RL/4H5 low profile base
YBF-RL/4 base[3]
YBC-RL/4H5 base
AMU-B2 addressable base
AMU-MB master addressable base
YBF-RL/3J slave to AMU-MB master addressable base
YBF-RL/2NBD analogue sensor base
YBC-R/3 base
YCA-RL/3H2 addressable base[3]
YCA-RL/5H2 addressable master base[3]
YBJ-RL/2NA analogue sensor base
YBK-RL/4H1A base[3]
YBK-RL/4H2 base[3]
YBN-R/2NA addressable base
YBN-R/4 base
YBN-R/3[3] base
YBO-R/4C base
YBN-R/4C base

[2] Detector/base combinations also meet the requirements of the standard in low power mode.
[3] *Base manufactured by Hochiki Europe.*

Hochiki Europe (U.K.) Limited

Grosvenor Road, Gillingham Business Park, Gillingham, Kent ME8 0SA

Tel: +44 (0)1634 260131 • Fax: +44 (0)1634 260132

Certificate No: 164a to BS 5445: Part 7
Point smoke detectors

SLK-E photoelectric smoke detector (YBF-RL/4H5[2], YBF-RL/4[2], AMU-B2, AMU-MB, YBF-RL/3J, YBC-R/3, YCA-RL/3H2[1,2], YCA-RL/5H2[1,2], YBK-RL/4H1A[2] and YBK-RL/4H2[2] bases)	164a/01
SIH-E ionisation smoke detector (YBF-RL/4H5[2], YBF-RL/4[2], AMU-B2, AMU-MB, YBF-RL/3J, YBC-R/3, YCA-RL/3H2[1,2], YCA-RL/5H2[1,2], YBK-RL/4H1A[2] and YBK-RL/4H2[2] bases)	164a/02
SLK-ED integrating photoelectric smoke detector (YBF-RL/4H5, AMU-B2, AMU-MB, YBF-RL/3J, YBC-R/3, YCA-RL/3H2[1,2], YCA-RL/5H2[1,2], YBK-RL/4H1A[2] and YBK-RL/4H2[2] bases)	164a/03
SLK-EN conventional photoelectric smoke detector (YBF-RL/4H3H, YBK-RL/4H2, YBK-RL/4H1A, YBF-RL/4H5, YCA-RL/5H2, YCA-RL/3H2 bases)	164a/05
SLR-E conventional photoelectric smoke detector (YBN-R/4 bases)	164a/06

Bases

YBF-RL/4H5[2] low profile base	164/a04

YBF-RL/4 base[2]
AMU-B2 addressable base
AMU-MB master addressable base
YBF-RL/3J slave to AMU-MB master addressable base
YBC-R/3 base
YCA-RL/3H2[2] addressable base
YCA-RL/5H2[2] addressable master base
YBK-RL/4H1A base[2]
YBK-RL/4H2 base[2]
YBJ-RL/2NA analogue sensor base [2, 3]
YBF-RL/4H3H[2]

Notes:
1 Detector/base combinations also meet the requirements of the standard in low power mode
2 Bases manufactured by Hochiki Corporation, except for [2].
3 See Hochiki Corporation listing for details of compatible detectors.

Honeywell Incorporated

8500 Bluewater Road, NW Albuquerque, New Mexico, 87121-1958, USA

Tel: +1 505 831 7000 • Fax: +1 505 831 7420
UK Sales Enquiries:
Honeywell Control Systems Limited
Tel: +44 (0)1344 656000 • Fax: +44 (0)1344 656240

Certificate No: 118a to BS 5445: Part 5
            118b to BS 5445: Part 7

Certificated Products	**LPCB Ref. No.**
Point smoke detectors	
TC807A1036 analogue ionisation smoke sensor (14506414-002 base)	118b/03
TC806A1037 analogue photoelectric smoke sensor (14506414-002 base)	118b/04
TC804E1030 conventional photoelectric smoke detector (14506587-005 base)	199b/05
TC805E1013 conventional ionisation detector (14506587-005 base).	199b/08
TC807E1003 analogue ionisation smoke detector (14506414-007 base)	199b/03
TC806E1020 analogue photoelectric smoke sensor (14506414-007 base)	199b/04
TC806E1012 analogue photoelectric smoke sensor (14506414-007 base)	199b/06

Note: Meets BS 5445: Part 7 (EN54 Part 7) with detector 'Low', 'Normal' and 'High' sensitivity setting.

TC807E1011 analogue ionisation smoke detector (14506414-007 base)	199b/07

Point heat detectors

TC808A1043 grade 1 analogue heat sensor (14506414-002 base)	118a/02
TC808E1028 grade 1 analogue heat sensor (14506414-007 base)	199a/02
TC830E1004 grade 1 rate of rise heat detector (14506587-005 base)	199a/03
TC808E1002 grade 2 fixed temperature heat detector (14506414-007 Base)	199a/05

Bases
14506414-002 base
14506414-007 analogue detector base
14506587-005 conventional resistor base

Kidde Fire Protection Limited
12 Nelson Industrial Estate, Cramlington, Northumberland NE23 9BL
Tel: +44 (0)1670 713455 • Fax: +44 (0)1670 735553

Quality System Cert. No. 054 Assessed to ISO 9001 : 1994

Approved Products	LPC Ref. No.
HART aspirating high-sensitivity (SMART variant) smoke detector	054a/01

The Approval is conditional on;
1. Commissioning smoke tests are performed on all installations in accordance with Kidde Hartnell Installation and Commissioning Manuals.
2. Installations are wired using screened cable only.
3. Control panels using the product are designed to comply with the Kidde Hartnell Engineering Document 0036, entitled 'Design Specification for (Smart) Detector'.

Menvier Limited
Wildmere Road, Banbury, Oxon OX16 7TU
Tel: +44 (0)1295 270100 • Fax: +44 (0)1295 268972

Certificate No: 250a to BS 5445: Part 7
250b to BS 5445: Part 5 and 8

LPCB Ref No:

Point smoke detectors
MPD720 Conventional Photoelectric Smoke Detector (MDB700 base)	250a/01
MAI710S analogue ionisation smoke sensor (MDB700 base)	250a/02
Note: Meets the requirements of BS5445: Part 7 at normal sensitivity setting only.	
MAP720S analogue photoelectric smoke sensor (MDB700 base)	250a/03
Note: Meets the requirements of BS5445: Part 7 at normal sensitivity setting only.	

Point heat detectors
MAH730S analogue heat sensor (MDB700 base)	250b/01
Note: Meets BS5445: Part 5, grade 1 or 2 and BS5445: Part 8, range 2	

Nittan (UK) Limited
Hipley Street, Old Woking, Surrey GU22 9LQ
Tel: +44 (0)1483 769555/8 • Fax: +44 (0)1483 756686

Certificate No: 041a to BS 5445: Part 5
041b to BS 5445: Part 8
041c to BS 5445: Part 7
367a to BS5445 : Part 7

Certificated Products — **LPCB Ref. No.**
Point smoke detectors
2KH-4T photoelectric smoke detector (RB4 and RB-3R bases)	041c/01
2IC-4T ionisation smoke detector (RB4 and RB-3R bases)	041c/02
2KC-AS-2LR analogue photoelectric smoke sensor (RB4 and RB-3R bases)	041c/04
Note: Meets BS 5445: Part 7 (EN54: Part 7) with 'normal' and 'high' sensitivity settings only	
ST-I conventional ionisation smoke detector (STB-4)	041c/05
ST-P conventional photoelectric smoke detector (STB-4)	041c/06
ST-I-P conventional ionisation smoke detector (STB-4 base)	041c/07
ST-P-P conventional photoelectric smoke detector (STB-4 base)	041c/08
ST-I conventional ionisation smoke detector (STB-4 base)	367a/01
ST-P conventional photoelectric smoke detector (STB-4 base)	367a/02
2000/ION conventional ionisation smoke detector (STB-3 base)	367a/03
2000/OP conventional photoelectric smoke detector (STB-3 base)	367a/04
ST-P-AS analogue photoelectric smoke sensor (STB-4 base)	041c/09
Note: Meets BS 5445 Part 7 at low, normal and high sensitivity settings.	
ST-I-AS analogue ionisation smoke sensor (STB-4 base)	041c/10
Note: Meets BS 5445 Part 7 at low, normal and high sensitivity settings.	

Continued

Nittan (UK) Limited (continued)

Point heat detectors
NHD-G1(4T) grade 1 fixed temperature heat detector (RB4 and RB-3R bases)   041a/01
NHD-G2(4T) grade 2 fixed temperature heat detector (RB4 and RB-3R bases)   041a/02
NHD-G3(4T) grade 3 fixed temperature heat detector (RB4 and RB-3R bases)   041a/03
TCA-AS-2LR analogue heat sensor (RB4 base)   041a/04
   Note: Certificated when configured to response grade 1, 2 or 3 only
NHD-GH1(4T) range 1 fixed temperature heat detector (RB4 and RB-3R bases)   041b/01

LPCB
LOSS PREVENTION
CERTIFICATION BOARD

Approved Products   **LPC Ref. No.**
Point smoke detectors
NID 58-AW[(1)] analogue ionisation smoke sensor (3RB-5-AW base)   47/19
2KC-AW[(1)] analogue photoelectric smoke sensor (3RB-5-AW base)   47/22
Point heat detectors
TCA-AW[(1)] analogue heat sensor (3RB-5-AW base)   47/20
Bases
RB4 standard base
RB/3 standard base
RB/3R resistor base
RB6 standard base
3RB-5-AW[(1)] addressable base
STB-4 mounting base
(1) For exclusive use of Wormald Ansul

Nittan Electronic Company Limited
563-6 1-chome, Toyonodai, Otonemachii, Kitasaitama-gun, Saitama, Japan
Sales Enquiries: Nittan Co. Limited
Tel: +81 3 3468 1111 • Fax: +81 3 3468 4553
Certificate No: 069a to BS 5445: Part 7 (EN 54: Part 7)

Certificated products   **LPCB Ref. No.**
2KH1-LS photoelectric smoke detector (3T base)   069a/01
Bases
3T base

Nittan Seiki Company Limited
1, Aza Minamihata, Kohnoike, Itami, Hyougo 664, Japan
Sales Enquiries: Nittan Co. Limited
Tel: +81 3 3468 1111 • Fax: +81 3 3468 4553
Certificate No: 191a to BS 5445: Part 7 (EN 54: Part 7)

LPCB

Certificated products   **LPCB Ref. No.**
2IC-LS ionisation smoke detector (3T base)   191a/01
Bases
3T base

Notifier Limited
Charles Avenue, Burgess Hill, West Sussex RH15 9UF

Tel: +44 (0)1444 230300 • Fax: +44 (0)1444 230888

Certificate No: 118a to BS 5445: Part 5
118b to BS 5445: Part 7
199a to BS 5445: Part 5
199b to BS 5445: Part 7

Certificated Products	LPCB Ref. No.
Point smoke detectors	
CPX-551 analogue ionisation smoke sensor (B501 base)	118b/03
CPX-551E analogue ionisation smoke sensor (B501 base)	199b/03
SD-651E conventional photoelectric smoke detector (B401 base)	199b/05
SDX-751E analogue photoelectric smoke sensor (B501 base)	199b/06
Note: Meets BS 5445: Part 7 (EN54 Part 7) with detector 'Low', 'Normal' and 'High' sensitivity setting.	
SDX-551HRE analogue photoelectric smoke sensor (B501 base)	199b/04
CPX-751E analogue ionisation sensor (B501 base)	199b/07
CP-651E conventional ionisation detector (B401 base)	199b/08
SDX-551 analogue photoelectric smoke sensor (B501 base)	118b/04
Point smoke detectors	
LPX-751 VIEW Smoke Sensor (B501 base)	118b/10
Note: Meets BS 5445: Part 7 (EN 54 Part 7) in low sensitivity mode setting AL9=1%/f	
Point heat detectors	
FDX-551R grade 1 analogue heat sensor (B501 base)	118a/02
FDX-551E grade 2 analogue heat sensor (B501 base)	199a/05
FDX-551RE grade 1 analogue heat sensor (B501 base)	199a/02
Bases	
B501 base	
B401 base	

Photain Controls plc
Rudford Estate, Ford Aerodrome, Arundel, West Sussex BN18 0BE

Tel: +44 (0)1903 721531 • Fax: +44 (0)1903 726795

Certificate No: 066a to BS 5445: Part 7 (EN54: Part 7)
066b to BS 5445: Part 5 (EN54: Part 5)
066c to BS 5445: Part 8 (EN54: Part 8)

Certificated Products	LPCB Ref. No.
Point smoke detector	
ISD-P91 ionisation smoke detector (FSB-P91 base)	066a/01
ISD-I94 analogue ionisation smoke sensor (FSB-I94 base)	066a/02
Point heat detectors	
PRR-P91 grade 1 rate of rise heat detector (FSB-P91 base)	066b/01
PHD-P90 range 1 fixed temperature heat detector (FSB-P91 base)	066c/01
PRR-I94 Grade 3 analogue rate of rise heat sensor (FSB-194 base)	066b/02
Base	
FSB-P91 base	
FSB-I94 base	

Pittway Tecnologica SpA
Via Caboto 19, 34147 Trieste, Italy

European Sales Enquiries: System Sensor Europe
Tel: +44 (0)1403 276500 Fax: +44 (0)1403 276501

Certificate No: 199a to BS 5445: Part 5
199b to BS 5445: Part 7

Certificated Products	LPCB Ref. No.
Point smoke detector	
1551E analogue ionisation smoke sensor (B501 base)	199b/03
1251E analogue ionisation smoke sensor (B501 base)	199b/07

Continued

Pittway Technologica SpA (continued)

Note: Meets BS 5445: Part 7 (EN54: Part 7) with 'Low','Normal' and 'High' sensitivity settings.

1151E ionisation smoke detector (B401, B401R and B401SD bases)	199b/08
2551HRE analogue photoelectric smoke sensor (B501)	199b/04
2251E analogue photoelectric smoke sensor (B501 base)	199b/06

Note: Meets BS 5445: Part 7 (EN54:Part 7) with low, normal and high sensitivity settings.

2151E Conventional Photoelectric Smoke Detector (B401 base)	199b/05

Point heat detectors

5551RE grade 1 analogue rate of rise heat sensor (B501 base)	199a/02
5451E grade 1 rate of rise heat detector (B401, B401R and B401SD bases)	199a/03
5551E grade 2 analogue heat sensor (B501 base)	199a/05
5551 HTE range 1 analogue high temperature heat detector (B501 base)	199f/01

Bases
B401 conventional detector base
B501 analogue sensor base
B401R conventional resistor base
B401SD conventional Scholtky check base

Rafiki Protection Limited

Unit 55, Springvale Industrial Estate, Cwmbran, Gwent NP44 5BD

Tel: +44 (0)1633 865558 • Fax: +44 (0)1633 866656

Certificate No: 331a to BS 5445 Parts 5 and 7

Certificated products	**LPCB Ref. No.**
Multipoint combined heat and smoke detector (26-001 base)	331a/01

Notes: Certificated in modes 'Heat 1' and 'Smoke 2' only. Heat 1 meets the requirements of BS 5445: Part 5 grade 1 and Smoke 2 meets the requirements of BS 5445: Part 7

Siemens Building Technologies AG, Cerberus Division

Alte Landstrasse 411, CH-8708 Männedorf, Switzerland.
Manufacturing at: Siemens Building Technologies AG, Cerberus Division, Volketswil Factory, Industriestrasse 22, CH-8604 Volketswil, Switzerland

Tel: +41 1 922 6111 • Fax: +41 1 922 6450
UK Sales: Cerberus Limited
Tel: +44 (0)1734 783703 • Fax: +44 (0)1734 775750

Certificate No: 126a to BS 5445: Part 7
126b to BS 5445: Part 5
126c to BS 5445: Part 8
126e to BS 5445: Parts 5 and 8
127a to BS 5445: Part 7 (Issued to Nelm AG)
126k to BS5839: Part 5

Certificated products	**LPCB Ref. No.**
Point smoke detectors	
DO1131A photoelectric smoke detector, addressable (DB1131A base)	126f/01
F726 ionisation smoke detector (Z74 and Z74A bases)	127a/01
DO1101A-Ex photoelectric smoke detector, conventional (DB1101A base)	126a/07
DO1101A photoelectric smoke detector, conventional (DB1101A base)	126a/08
DOT1131A Photoelectric smoke detector with thermal enhancement, addressable (DB1131A base)	126a/09

Notes: Meets the requirements of BS5445: Part 7 (EN54:Part 7) in normal and high sensitivity settings only. Detector incorporates line isolation facility.

DO1151A photoelectric smoke detector, addressable (DB1151A base)[1]	126a/10
DO1152A photoelectric smoke detector, addressable (DB1151A base)[1]	126a/11
DOT1151A photoelectric smoke detector with thermal enhancement, addressable (DB1151A base)[1]	126a/12
DOT1152A photoelectric smoke detector with thermal enhancement, addressable (DB1151A base)[1]	126a/13
DO1131A photoelectric smoke detector, addressable (DB1131A base)	126a/14

Note: Detector incorporates line isolation facility

Continued

3 SECTION 4:
DETECTORS

Siemens Building Technologies AG, Cerberus Division (continued)

Point heat detectors
DT1152A rate of rise heat detector addressable (DB1151A base)   126e/02
   Note: Meets BS 5445: Part 5: grade 1 and BS 5445: Part 8, range 1. Detector incorporates line isolation facility
DT1101A-Ex grade 1 rate of rise heat detector, conventional (DB1101A base)   126b/02
DT1101A grade 1 rate of rise heat detector, conventional (DB1101A base)   126b/03
DT1102A-Ex range 1 rate of rise high temperature heat detector, conventional (DB1101A base)   126c/02
DT1102A range 1 rate of rise high temperature heat detector, conventional (DB1101A base)   126c/03
   Note: Detector incorporates line isolation facility
DT1131A grade 1 rate of rise heat detector, addressable (DB1131A base)   126b/04
   Note: Detector incorporates line isolation facility

Beam detectors
DLO1191 optical beam smoke detector (DLB1191A base)   126k/01

LPCB
LOSS PREVENTION
CERTIFICATION BOARD

Approved Products   **LPC Ref. No.**
Point smoke detectors
F910 ionisation smoke detector (ZA6/9, Z74A, Z74, Z94, Z94I, Z94SI, Z94MI, Z94B, Z94D bases)   18/24
F906 ionisation smoke detector (ZA6/9, Z74A, Z74, Z94, Z94I, Z94SI, Z94MI, Z94B, Z94D bases)   18/27
F911 ionisation smoke detector (Z91C, ZA6/9Ex, Z94C bases)   18/31
R716I optical smoke detector (Z74A,Z74,Z94, Z94B, Z94D, Z94I,Z94SI,Z94MI bases)   18/70
F716I ionization smoke detector (Z74A,Z74,Z94, Z94B, Z94D, Z94I,Z94SI,Z94MI bases)   18/71
F930 ionisation smoke detector (Z94, Z94B, Z94D, Z94I, Z94SI, Z94MI bases)   18/76
R930 photoelectric smoke detector (Z94, Z94B, Z94D, Z94I, Z94SI, Z94MI bases)   18/77
Point heat detectors
D716 grade 2 rate of rise heat detector (Z74 and Z74A bases)   18/22
D900 grade 1 rate of rise heat detector (ZA6/9, Z74A, Z74, Z94, Z94I, Z94SI, Z94MI, Z94B, Z94D bases)   18/23
D920 range 1 rate of rise heat detector (ZA6/9, Z74A, Z74, Z94, Z94I, Z94SI, Z94MI, Z94B, Z94D bases)   18/25
D901 grade 1 rate of rise heat detector (Z91C, ZA6/9Ex, Z94C bases)   18/32
D921 range 1 rate of rise heat detector (Z91C, ZA6/9Ex, Z94C bases)   18/33
Flame detectors
S610([1]) infrared flame detector (FKS6.1 base)   18/19
S2406([2]) infrared flame detector   18/72
S2406Ex([2]) intrinsically safe infrared flame detector   18/73
(1) S610 detectors shall be installed in accordance with the 'Rules for Automatic Fire Alarm Installations for the Protection of
   Property' and also in accordance with the following special requirements;
    - The detector must be placed at the best possible location, i.e. looking straight down on to the protected area or object.
    - If objects (not entire areas) are protected, blinds and/or shielding must be used to keep stray light away from the sensor.
    - The sensitivity and/or response delay must be properly adjusted to the prevailing conditions at the place of installation.
    - There should be inspection and maintenance every 2-3 months, depending on the environmental conditions. If there is
    excessive dust or grease in the air, the optical lens must be cleaned.
(2) S2406 and S2406Ex detectors shall be installed in accordance with the Cerberus document x164c, dated December 1989.
Bases
DB1101 base for conventional addressing
DB1151 base for interactive fire detectors
DBZ1191 base for feeding in surface wiring
DBZ1192 base for use in damp environments
ZAS6 base
Z74A base
Z74 base
ZA6/9 base adapter
ZA6/9Ex base adapter
Z94 base
Z94I addressable base
Z94SI slave to master addressable base
Z94MI master addressable base
Z94B base
Z94D base
Z94C base
DB1131A base
DB1101A base
DB1131A base
DB1151A base

84

James Stuart (London) & Co. Limited

Detection House, Brooklands Approach, North Street, Romford, Essex RM1 1DX

Tel: +44 (0)1708 769314 • Fax: +44 (0)1708 725868

Certificate No:067a to BS 5445: Part 5
067b to BS 5445: Part 7

Certificated products	LPCB Ref. No.
GD94/1 grade 2 rate of rise heat detector (GD9 base)	067a/01
GD22 ionisation smoke detector (GDB base)	067b/02

Bases
GDB base
GD9 base

System Sensor

(A Division of Pittway), 3825 Ohio Avenue, Saint Charles, IL 60174, USA

Tel: +1 708 377 6580 • Fax: +1 708 377 6985
European Sales Enquiries: System Sensor Europe
Tel: +44 (0)1403 276500 • Fax: +44 (0)1403 276501

Certificate No: 118a to BS 5445: Part 5
118b to BS 5445: Part 7

Certificated Products	LPCB Ref. No.
Point Smoke Detectors	
1551 analogue ionisation smoke sensor (B501 base)	118b/03
2551 analogue photoelectric smoke sensor (B501 base)	118b/04
2151E photoelectric smoke detector (B401 base)	118b/05
2251 analogue photoelectric smoke sensor (B501 base)	118b/06
Note: Meets BS 5445: Part 7 (EN54: Part 7) with 'Low', 'Normal', and 'High' sensitivity settings.	
1251 analogue ionisation smoke sensor (B501 base)	118b/07
Note: Meets BS 5445: Part 7 (EN54: Part 7) with 'Low', 'Normal', and 'High' sensitivity settings.	
1151EIS intrinsically safe ionisation smoke sensor (B401 base).	118b/09
Point Heat Detectors	
5551R grade 1 analogue heat sensor (B501 base)	118a/02
5451EIS grade 1 rate of rise intrinsically safe heat detector (B401 base)	118a/04

Bases
B401 conventional detector base
B501 analogue sensor base

Approved Products

Multi criteria detectors	
3251 intelligent ionisation, optical and thermal fire sensor (B524IE and B501 bases)	118c/01
Notes : Meets the sensitivity requirements of EN54 : Part 7 at all 5 sensitivities.	

Thorn Security Limited

160 Billet Road, Walthamstow, London E17 5DR

Tel: +44 (0)181 919 4000 • Fax: +44 (0)181 919 4040

Certificate No: 143a to BS 5445: Part 7
Certificate No: 143e to BS 5445: Part 5 and BS 5445: Part 8
Certificate No: 143f to BS 5445: Part 5

Certificated Products	LPCB Ref. No.
Point Smoke Detectors	
MR301T High performance optical (HPO) smoke detector (M300 base)	143a/01
MR501T High performance optical (HPO) analogue smoke sensor (M500 base)	143a/02
Note: Meets BS 5445: Part 7 with normal sensitivity only	

Continued

3 SECTION 4:
DETECTORS

Thorn Security Limited (continued)

MF901 analogue addressable low profile ionisation smoke detector (M600/900 base)[1]	143a/03
Note: [1]Meets the requirements of BS 5445 : Part 7 at high and normal sensitivity settings only	
MR901 analogue addressable low profile photoelectric smoke detector (M600/900 base)[3]	143a/04
[3]Note: Meets the requirements of BS 5445 : Part 7 at HIGH and NORMAL sensitivity settings only.	
MR601 conventional low profile photoelectric smoke detector	
(M600/900 and M600 diode continuity base)	143a/05
MF601 conventional low profile ionisation smoke detector	
(M600/900 and M600 diode continuity base)	143a/06
MR601T conventional low profile high performance optical smoke detector	
(M600/900 and M600 diode continuity base)	143a/07
MR901T analogue low profile high performance optical smoke detector (M600/900 base)	143a/08
Note: Meets BS 5445 : Part 7 at High and Normal sensitivity settings only.	
Point Heat Detectors	
MD901 analogue low profile heat sensor (M600/900 base)	143e/01
Note: Meets BS 5445 : Part 5, grade 1, 2 or 3 and BS 5445 : Part 8, range 2.	
MD601 conventional grade 1 low profile heat detector. (M600/900 and M600 diode continuity base)	143f/01
MD611 conventional fixed temperature grade 2 low profile heat detector (M600/900 and M600 diode continuity base)	143f/02

LPCB

Approved Products	LPC Ref. No.
Point heat detectors	
H600 grade 3 fixed temperature heat detector	10/28
H601 grade 3 fixed temperature heat detector	10/29
H602 grade 3 fixed temperature heat detector	10/30
H900 range 2 fixed temperature heat detector	10/31
H901 range 2 fixed temperature heat detector	10/32
H902 range 2 fixed temperature heat detector	10/33
MD301 grade 1 rate of rise heat detector (M300 base)	10/65
MD311 grade 2 fixed temperature heat detector (M300 base)	10/66
MD401 addressable grade 1 rate of rise heat detector (M500 base)	10/73
MD501 analogue heat sensor (M500 base)	10/76
Note: Meets BS 5445: Part 5 (EN54: Part 5), grade 1, 2 or 3, and BS 5445: Part 8 (EN54: Part 8), range 2.	
Point Smoke Detectors	
MF301 ionisation smoke detector (M300 base)	10/63
MR301 photoelectric smoke detector (M300 base)	10/64
MF301D ionisation smoke detector (M300 base)	10/67
MF301DH ionisation smoke detector (M300 base)	10/68
MR401 addressable photoelectric smoke detector (M500 base)	10/74
MF401 addressable ionisation smoke detector (M500 base)	10/75
MR501 analogue photoelectric smoke sensor (M500 base)	10/77
Note: Meets BS 5445: Part 7 (EN54: Part 7) with 'Normal' and 'High' sensitivity settings only.	
MF501 analogue ionisation smoke sensor (M500 base)	10/78
Flame Detectors	
S111 infrared flame detector[(1)]	10/25
S112 infrared flame detector[(1)]	10/26
S121 infrared flame detector[(1)]	10/27
MS502Ex intrinsically safe flame detector (M500Ex base)	10/103
S161 infrared flame detector [(2)]	143d/01

(1) S111 range flame detectors shall be installed in accordance with Thorn Security publication 'User Manual IR-111,112 and 121'.

(2) S161 flame detectors shall be installed in accordance with Thorn Security publication 'S161 Installation Guide: 01A-04-I5'.

Bases
M300 standard base
M500 addressable base
M500Ex intrinsically safe addressable base
M600/900 base
M600 diode continuity base

Transmould Limited
Ballyspillane Industrial Estate, Killarney, County Kerry, Ireland

Tel: +353 64 34158 • Fax: +353 64 34161
UK Sales Enquiries: Menvier (Electronic Engineers)
Tel: +44 (0)1295 256363 • Fax: +44 (0)1295 270102

Certificate No: 180a to BS 5445: Part 7
180b to BS 5445: Part 5
180c to BS 5445: Part 8

Certificated Products	LPCB Ref. No.
Point smoke detectors	
MID 710 (Menvier) ionisation smoke detector (MDB700 base)	180a/01
FSID 10 (Transmould) ionisation smoke detector (MDB700 base)	180a/01
TS7502A (Tann Synchronome) ionisation smoke detector (MDB700 base)	180a/01
MPD 720[1] (Menvier) photoelectric smoke detector (MDB700 base)	180a/02
FSPD 20[1] (Transmould) photoelectric smoke detector (MDB700 base)	180a/02
TS7501A[1] (Tann Synchronome) photoelectric smoke detector (MDB700 base)	180a/02
Note: [1]Both ASIC and hybrid versions are certificated.	
Point heat detectors	
MFR 730 (Menvier) grade 1 rate of rise heat detector (MDB700 base)	180b/01
FSFR 30 (Transmould) grade 1 rate of rise heat detector (MDB700 base)	180b/01
TS7503A (Tann Synchronome) grade 1 rate of rise heat detector (MDB700 base)	180b/01
MMT 760 (Menvier) grade 2 rate of rise heat detector (MDB700 base)	180b/02
FSMT 60 (Transmould) grade 2 rate of rise heat detector (MDB700 base)	180b/02
TS7504A (Tann Synchronome) grade 2 rate of rise heat detector (MDB700 base)	180b/02
MHT 790 (Menvier) range 2 rate of rise heat detector (MDB700 base)	180c/01
FSHT 90 (Transmould) range 2 rate of rise heat detector (MDB700 base)	180c/01
TS7505A (Tann Synchronome) range 2 rate of rise heat detector (MDB700 base)	180c/01
Bases	
MDB700 base	

Vision Products Pty Limited
15-17 Normanby Road, Clayton, Victoria 3168, Australia

Tel: +61 3 954 48411 • Fax: +61 3 954 48648
European Sales & Support: Vision Systems (Europe) Ltd
Tel: +44 (0)1442 242330 • Fax: +44 (0)1442 249327

Certificate No: 305b to GEI 1-048 30-01-97

Approved Products	LPCB Ref. No.
VESDA E70-D EMC high sensitivity, aspirating smoke detector(1)	305a/01

(1) The approval is conditional on the following requirements:
- Commissioning tests are performed on all installations in accordance with VESDA Design and Applications Manual by installers accredited by the manufacturer.
- Periodicity of detector calibration shall not exceed 3 years.
- All requirements and notes highlighted in Vision Systems Installation and Programming Manual shall be complied with.
- Each area protected contains a minimum of 2 sampling holes.

VESDA LaserPLUS high sensitivity aspirating smoke detector   305b/01
Approved configurations:
VLP-012 Detector with Display & Programmer
VLP-002 Detector with Display
VLP-400 Detector with Fire/OK LEDs only
VLP-401 Detector with Programmer & Fire/OK LEDs

Approved remote units
VRT-100 Programmer
VRT-200 Display unit (with relays)
VRT-300 VESDAnet socket
VRT-500 Relay unit
VRT-600 Display unit (without relays)
Notes: For compliance with clause 4.13 of GEI-1-048 the LaserPLUS detector shall be supplied with power from a power supply complying with the requirements of EN54 Part 4.
The approval is conditional upon the following requirements:
- Design, Installation and Commissioning are performed in accordance with VESDA System Design Manual by installers accredited by the manufacturer.
- Each area is protected by a minimum of 2 holes.

3 SECTION 4:
DETECTORS

Wormald Ansul (UK) Limited - Wormald Engineering
Wormald Park, Grimshaw Lane, Newton Heath, Manchester M40 2WL

Tel: +44 (0)161 205 2321 • Fax: +44 (0)161 455 4459

Approved Products	**LPC Ref. No.**
IR6003 optical beam smoke detector[1]	60/17
Universal Interface Module (U.I.M.)	60/18

(1) IR6003 detectors shall be installed in accordance with the following document:
WORMALD I/R BEAM SMOKE/PARTICLE DETECTOR, MODEL I/R6003, Revision 'D' - 21 DECEMBER 1989

Ziton SA (Pty) Limited
Ziton House, 9 Buitenkant Street, Cape Town 8000, South Africa

UK Sales enquiries: Ziton Limited
Tel: +44 (0)1908 281981 • Fax: +44 (0)1908 282554

Certificate No: 092a to BS 5445: Part 5
092b to BS 5445: Part 8
092c to BS 5445: Part 7
092e to BS 5445: Parts 5 and 7

Certificated Products

Point smoke detectors

	LPCB Ref. No.
Z610-1 (fixed sensitivity) ionisation smoke detector[1]	092c/01
Z610A-1 (adjustable sensitivity) ionisation smoke detector[1][2]	092c/02
Z630-1 (fixed sensitivity) optical smoke detector[1]	092c/03
Z630A-1 (adjustable sensitivity) optical smoke detector[1][2]	092c/04
ZP710-2 analogue ionisation smoke sensor[3][4]	092c/05
ZN 730-2 analogue optical smoke detector [4][6]	092c/06
ZP730-2 analogue optical Smoke Sensor [2,3]	092c/07

Point heat detectors

Z620-582-1 grade 1 rate of rise heat detector[1]	092a/01
Z620-821-1 range 1 fixed high temperature heat detector[1]	092b/01
Z620-581-1 grade 2 fixed temperature detector[1]	092a/02
ZP720-2 grade 1 analogue heat sensor[3][5]	092a/03
ZN 720-2 grade 1 analogue heat sensor [6]	092a/04

Point smoke/heat detectors

ZN832-2 analogue addressable optical smoke heat detector (ZG BS1 base)	092e/01

Meets: BS 5445 Part 7 as optical only at low and normal sensitivities and as a thermally enhanced optical detector.
In addition BS 5445 Part 5 grades 1, 2 and 3

(1) Certificated with Ziton Z6-BS1, Z6-BS2 and Z6-BS4 bases.
(2) Meets BS 5445: Part 7 (EN54: Part 7) at all selectable sensitivity settings.
(3) Certificated with Ziton ZP7-SB1 base.
(4) Meets BS 5445: Part 7 (EN54: Part 7) at normal sensitivity only.
(5) Meets BS 5445: Part 5 (EN54: Part 5) at sensitivity state 2 only.
(6) Certificated with the ZG-BS1 base

Bases
Z6-BS1 surface mounting base
Z6-BS2 surface mounting base, Schottky diode
Z6-BS4 surface mounting base, diode
ZP7-SB1 surface base

INTRODUCTION
Manual call points are assessed to the requirements of : BS 5839 Part 2 - Specification for manual call points.

Audit testing
Once listed, a manual call point is eligible for audit testing to ensure continued compliance with BS 5839: Part 2.

Note
The LPCB approves and certificates products to the requirements of product standards. Where practical, the LPCB, in consultation with the manufacturer, will also verify that the products can meet the recommendations of applicable codes of practice.

The LPCB uses national and international standards for the listing of products. In some instances the requirements of these standards may conflict with the recommendations of local codes of practice. When this situation arises, it is on the requirements of the product standard that the LPCB base approval and certification.

If full compliance with an installation code of practice is necessary, confirmation should be obtained that the listed products also comply with the appropriate codes of practice.

Apollo Fire Detectors Limited
36 Brookside Road, Havant, Hampshire PO9 1JR

Tel: +44 (0)1705 492412 • Fax: +44 (0)1705 492754

Certificate No: 010d to BS 5839: Part 2

Certificated Products	LPCB Ref. No.
54000-901 addressable manual call point	010d/01
55000-910 addressable manual call point	010d/02
Note: Certificated with Series 90 and XP95 communication protocols	
55000-940 intrinsically safe manual call point[1]	010d/03
Note: Certificated with Series 90 and XP95 communication protocols.	
55000-950 Waterproof addressable manual callpoint	010d/04
Note: Certificated with Series 90 and XP95 communication protocols	

1 The LPCB certification of these devices does not include the electrical parameters and marking concerning INTRINSIC SAFETY (I.S.) - users should confirm with local regulatory bodies and by consultation with the manufacturer that the devices are separately and correctly IS certificated for their application.

Caradon Gent Limited
140 Waterside Road, Hamilton Industrial Park, Leicester LE5 1TN

Tel: +44 (0)116 2462000 • Fax: +44 (0)116 2462300

Certificate No: 042c to BS 5839: Part 2

Certificated Products	LPCB Ref. No.
1195 OR flush and surface mounted manual call point	042c/01
13480-02 addressable manual call point	042c/02
34800 flush and surface mounted addressable manual call point	042c/03
Note: Approved for flush mounting when using 13480-29 flush mounting plate	
32800 flush and surface mounted addressable manual call point	042c/04
Note: Approved for flush mounting when using 19289-01 flush mounting plate.	
78150-52NM flush and surface mounted addressable manual call point.	042c/05
Note: Approved for flush mounting when using 78150-98NM flush mounting plate	

Middle East Sales Enquiries: MK Middle East Marketing Office, Dubai
Tel: +971 4 668 483 • Fax: +971 4 681 585

India Sales Enquiries: MK India, Madras
Tel: +91 44 626 9991 • Fax: +91 44 626 9992

Far East Sales Enquiries: MK Electric (Singapore) Pte. Limited
Tel: +65 271 7266 • Fax: +65 274 9219

3 SECTION 5:
MANUAL CALL POINTS

Cerberus AG

Alte Landstrasse 411, CH-8708 Männedorf, Switzerland

Tel: +41 1 922 6111 • Fax: +41 1 922 6450
UK Sales: Cerberus Limited: Tel 01734 783703 • Fax: +44 (0)1734 775750

For listing of LPCB certificated and approved products, please refer to new entry under **'Siemens Building Technologies AG, Cerberus Division'.**

Fulleon Synchrobell Limited

40 Springvale Industrial Estate, Cwmbran, Gwent NP44 5BD

Tel: +44 (0)1633 872131 • Fax: +44 (0)1633 866346

Certificate No. 378a to BS 5839 : Part 2
Certificated Products   **LPCB Ref. No.**
RC/S/R indoor manual call point, flush mounted   378a/01
Accessories
/BB surface mounted (plastic)

GLT Exports Limited

Detection House, 72-78 Morfa Road, Swansea SA1 2EN

Tel: +44 (0)1792 455175 • Fax: +44 (0)1792 455176

Certificate No. 330c to BS 5839: Part 2

Certificated Products   **LPCB Ref. No.**
ZT/MCP Conventional manual call point   330c/01
ZT-MCP/AD addressable manual call point   330c/02

Honeywell Incorporated

8500 Bluewater Road, NW Albuquerque, New Mexico, 87121-1958, USA

Tel: +1 505 831 7000 • Fax: +1 505 831 7535
UK Sales Enquiries: Honeywell Control Systems Limited

Tel: +44 (0)1344 656000 • Fax: +44 (0)1344 656240

Certificate No: 199c to BS5839: Part 2

Certificated Products   **LPCB Ref No:**
Honeywell SSDH500KAC addressable manual call point   199c/01

KAC Alarm Company Limited

KAC House, Tything Road, Arden Forest Industrial Estate, Alcester, Warwickshire B49 6EP

Tel: +44 (0)1789 763338 • Fax: +44 (0)1789 400027

Certificate No: 166a to BS 5839: Part 2

Certificated products
KR1 indoor manual call point, flush mounted (single pole changeover)   166a/01
KR72 indoor manual call point, flush mounted with series monitoring resistor   166a/02
KSR1 outdoor manual call point (single pole changeover)   166a/03
KSR72 outdoor manual call point with series monitoring resistor   166a/04

World Series Manual Call Points
WR2001 indoor manual call point, flush mounted (single pole changeover)   166a/06
WR2012 indoor manual call point, flush mounted   166a/07

Continued

KAC Alarm Company Limited (continued)

WR2004 indoor manual call point, flush mounted   166a/08
WR2061 indoor manual call point, flush mounted   166a/09
WR2072 indoor manual call point, flush mounted   166a/10
WR3001 indoor manual call point, flush mounted (single pole changeover)   166a/11
WR4001 outdoor manual call point (single pole changeover)   166a/12
WR4072 outdoor manual call point with series monitoring resistor   166a/13
WR2101 indoor manual call point (single pole changeover)   166a/14
WR3061 indoor manual call point, flush mounted   166a/15
WR3072 indoor manual call point, flush mounted   166a/16
WR3101 indoor manual call point, flush mounted (single pole changeover)   166a/17
WR4061 outdoor manual call point series and LED shunt resistor   166a/18
WRZ2/4001 outdoor manual call point (single pole changeover)   166a/19
WRZ2/4002 outdoor manual call point (end-of-line resistor)   166a/20
WRZ2/4010 outdoor manual call point (series monitoring and end-of-line resistors)   166a/21
WRZ2/4072 outdoor manual call point (series monitoring resistor)   166a/22
WRD/4001 outdoor manual call point (single pole changeover)   166a/23
WRD/4002 outdoor manual call point (end-of-line resistor)   166a/24
WRD/4010 outdoor manual call point (series monitoring and end-of-line resistors)   166a/25
WRD/4072 outdoor manual call point (series monitoring resistor)   166a/26
WR3631 indoor manual call point (SR3T-P mounting box)   166a/27
WR4061 R1-R2 (2.5W resistors) outdoor manual call point   166a/28
WR4031 outdoor manual call point (0.6 W resistors)   166a/29

Accessories for indoor manual call points
/SR   Surface mounted (plastic mounting box)
/SR2T-P   Surface mounted (plastic mounting box)[1]
/SR3T-P   Surface mounted (plastic mounting box)[1]
/SR4T-P   Surface mounted (plastic mounting box)[1]
/MR   Surface mounted (metal mounting box)
/ETT   European terminal tray ETT1[1] & ETT2[1] and ETT3[2]
/G   Grip forks[2]
/L   Earth continuity link (KL1)
BZR/1   Bezel[2]
BZR/2   Bezel Type 2[2]
M141   Spacer Piece[2]
BZR/3 Bezel Type 3[2]

(1) For use with World Series class 3000 products
(2) For use with World Series

World Series Manual Call points are approved with the following switch combinations:

	Otehall 382430	Otehall 95/89S	Otehall 382431	Honeywell V5T010TB3X144	Honeywell V5T020TBX145	Burgess XG3	Burgess GA43
WR2001	■			■		■	
WR2101		■	■		■		■
WR2004	■		■	■		■	
WR2012		■	■		■		■
WR2061		■	■		■		■
WR2072		■	■		■		■
WR3001	■			■		■	
WR3061		■	■		■		■
WR3072		■	■		■		■
WR3101		■	■		■		■
WR3631		■	■		■		■

Manual Call points are approved in the following materials:
Cycoloy C2800
Bayblend FR110
Bayblend FR1441
Noryl SE100
Cycoloy 2950

3 SECTION 5:
MANUAL CALL POINTS

Nittan (UK) Limited
Hipley Street, Old Woking, Surrey GU22 9LQ

Tel: +44 (0)1483 769555/8 • Fax: +44 (0)1483 756686

Certificate No: 041d to BS 5839: Part 2

Certificated Products	LPCB Ref. No.
NCP-AS-2LU addressable manual call point	041d/01

Pittway Tecnologica SpA,
Via Caboto 19, 34147 Trieste, Italy

European Sales Enquiries: System Sensor Europe
Tel: +44 (0)1403 276500 Fax: +44 (0)1403 276501

Certificate No: 199c to BS 5839: Part 2

Certificated Products

LPCB Ref. No.

M500KAC addressable manual call point — 199c/01

Approved with the following accessories:

/SR3T-P	Surface mounting box
/ETT	European terminal tray (ETT1 and ETT2)
/L	Earth continuity link KL1
BZR1	Bezel
BZR2	Bezel type 2
M141	Spacer piece

Siemens Building Technologies AG, Cerberus Division
Alte Landstrasse 411, CH-8708 Männedorf, Switzerland.
Manufacturing at: Siemens Building Technologies AG, Cerberus Division, Volketswil
Factory, Industriestrasse 22, CH-8604 Volketswil, Switzerland

Tel: +41 1 922 6111 • Fax: +41 1 922 6450
UK Sales: Cerberus Limited
Tel: +44 (0)1734 783703 • Fax: +44 (0)1734 775750

Certificate No: 126d to BS 5839: Part 2

Certificated Products	LPCB Ref. No.
DM1151 addressable manual call point	126d/01
DM1101 Indoor manual call-point (DMZ 1191 mounting box).	166a/05
DM1131 addressable manual call point	126d/02

LPCB

Approved Products	LPC Ref. No.
ATAN 50 manual call point	18/34

Signature Industries Limited - Clifford and Snell Alarms
Tom Cribb Road, Thamesmead, London SE28 0BH

Tel: +44 (0)181 316 4477/317 1717 • Fax: +44 (0)181 854 5149

LPCB

Approved Products	LPC Ref. No.
MCP/R/F manual call point	20/1
MCP/R/S manual call point	20/2

Thorn Security Limited

160 Billet Road, Walthamstow, London E17 5DR

Tel: +44 (0)181 919 4000 • Fax : 0181 919 4040

Certificate No: 143g to BS5839: Part 2

Certificated Products	LPCB Ref. No.
CP920 Addressable flush and surface mounted manual callpoint	143g/01
CP930 Addressable weatherproof manual callpoint	143g/02
CP200 Indoor manual call point	166a/09
CP210 Indoor manual call point	166a/09
CP230 Outdoor manual call point	166a/18
CP220Ex Outdoor manual call point	166a/28
CP540Ex Outdoor manual call point (0.6W resistors)	166a/29

Ziton (SA) Pty Limited

Ziton House, 9 Buitenkant Street, Cape Town 8000, South Africa

UK Sales Enquiries: Ziton Limited
Tel: +44 (0)1908 281981 • Fax: +44 (0)1908 282554

LPCB Ref. No.

	LPCB Ref. No.
ZP785-2 addressable flush and surface mounted manual call point	092d/01

INTRODUCTION

A line unit is a device which may be connected to a detection circuit or other transmission path of a fire detection system, used for functions other than detection such as;

- to receive or transmit information in relation to fire detection
- to provide functions, necessary for the operation of a fire detection system
- to provide functions for the control of fire protection systems.

For example, the following devices which may be connected to of a fire detection system,

are considered as line units:

- input/output interfaces
- short circuit isolators.

Products are currently tested to requirements obtainable through the LPCB.

Audit testing

Once listed, line modules become eligible for audit testing to ensure continued compliance with the applicable product standard.

Note

The LPCB approves and certificates products to the requirements of product standards.

Where practical, the LPCB, in consultation with the manufacturer, will also verify that the products can meet the recommendations of applicable codes of practice.

The LPCB uses national and international standards for the listing of products. In some instances the requirements of these standards may conflict with the recommendations of local codes of practice. When this situation arises, it is on the requirements of the product standard that the LPCB base approval and certification.

If full compliance with an installation code of practice is necessary, confirmation should be obtained that the listed products also comply with the appropriate codes of practice.

Apollo Fire Detectors Limited

36 Brookside Road, Havant, Hampshire PO9 1JR

Tel: +44 (0)1705 492412 • Fax: +44 (0)1705 492754

Certificate No: 010f to EFSG/F/95/007

Approved Products	**LPC Ref. No.**
54000-010 Series 90 short circuit isolator (45681-007 base)	13/20
55000-700 XP95 short circuit isolator (45681-211 base)	010e/01
55000-855 Single Channel Protocol Translator	010f/01
55000-856 Dual Channel Protocol Translator	010f/02

Bases
45681-211 short circuit isolator mounting base
45681-007 series 20/30 mounting base

Cerberus AG
Alte Landstrasse 411, CH-8708 Männedorf, Switzerland

Tel: +41 1 922 6111 • Fax: +41 1 922 6450
UK Sales: Cerberus Limited: Tel: +44 (0)1734 783703 • Fax: +44 (0)1734 775750

For listing of LPCB certificated and approved products, please refer to new entry under 'Siemens Building Technologies AG, Cerberus Division'.

Edwards System Technology
6411 Parkland Drive, Sarasota, Florida 34243, USA
Trading through ADT Security Systems for UK and Europe

Tel: +44 (0)171 407 9741 • Fax: +44 (0)171 407 1693

Approved Products	LPCB Ref. No.
M500MF monitor module	69/4
M500CFS control module	69/5
M500XF short circuit isolator	69/6
M501MF mini monitor module	69/18

Hochiki Europe (UK) Limited
Grosvenor Road, Gillingham Business Park, Gillingham, Kent ME8 0SA

Tel: +44 (0)1634 260131 • Fax: +44 (0)1634 260132

Certificate No: 164b to EFSG/F/95/007

Approved Products	LPCB Ref No.
CHQ-B dual sounder controller	164b/01
CHQ-MZ mini zone monitor	164b/02
CHQ-R dual relay controller	164b/03
CHQ-S dual switch monitor	164b/04
CHQ-Z dual zone monitor	164b/05
CHQ-SCI-S short circuit isolator	164b/06

Honeywell Incorporated
8500 Bluewater Road, NW Albuquerque, New Mexico, 87121-1958, USA

Tel: +1 505 831 7000 • Fax: +1 505 831 7535
UK Sales Enquiries: Honeywell Control Systems Limited
Tel: +44 (0)1344 656000 • Fax: +44 (0)1344 656240

Approved Products	LPC Ref. No.
TC809A1059 monitor module	69/4
TC810A1056 control module	69/5
TC811A1006 short circuit isolator	69/6
TC809B1008 mini monitor module	69/18
TC809E1019 monitor module	199d/01
TC810E1008 control module	199d/02
TC809E1027 mini monitor module	199d/04
TC811E1007 short circuit isolator	199d/03

3

SECTION 6:
LINE UNITS

Notifier Limited
Charles Avenue, Burgess Hill, West Sussex RH15 9UF

Tel: +44 (0)1444 230300 • Fax: +44 (0)1444 230888

Approved Products	LPC Ref. No.
MMX-1 monitor module	69/4
CMX-2 control module	69/5
ISO-X short circuit isolator	69/6
MMX-1E monitor module	199d/01
CMX-2E control module	199d/02
MMX-101E mini monitor module	199d/04
ISO-XE short circuit isolator	199d/03
MMX-101 mini monitor module	69/18

Pittway Tecnologica SpA
Via Caboto 19, 34147 Trieste, Italy

European Sales Enquiries: System Sensor Europe
Tel: +44 (0)1403 276500 Fax: +44 (0)1403 276501

Certificate No. 199e to EFSG/F/95/007.

Approved Products	LPCB Ref. No.
M500ME monitor module	199d/01
M500CHE control module	199d/02
M500XE short circuit isolator	199d/03
M501ME mini monitor module	199d/04
M503ME micromonitor module	199e/01

Protec Fire Detection plc
Protec House, Churchill Way, Nelson, Lancashire BB9 6RT

Tel: +44 (0)1282 692621 • Fax: +44 (0)1282 602570

Approved Products	LPC Ref. No.
AN/AD Mark 4 short circuit isolator.	201c/01

Siemens Building Technologies AG, Cerberus Division
Alte Landstrasse 411, CH-8708 Männedorf, Switzerland
Manufacturing at: Siemens Building Technologies AG, Cerberus Division, Volketswil
Factory, Industriestrasse 22, CH-8604 Volketswil, Switzerland

Tel: +41 1 922 6111 • Fax: +41 1 922 6450
UK Sales: Cerberus Limited: Tel: +44 (0)1734 783703 • Fax: +44 (0)1734 775750

Certificate No. 126j to EFSG/F/94/002
Certificate Number 126m to EFSG/F/95/007

Certificated Products	LPCB Ref.
DC1131 input module	126m/01
DC1134 output module	126m/02
DC1157 input module	126m/03

Approved Products	LPC Ref. No.
E90 MI master element	18/68
M5M 010 Multimaster	18/74
DC1151 input module	126j/01
DC1154 output module	126j/02

System Sensor

(A Division of Pittway), 3825 Ohio Avenue, Saint Charles, IL 60174, USA

Tel: +1 708 377 6580 • Fax: +1 708 377 6985
European Sales Enquiries: System Sensor Europe
Tel: +44 (0)1403 276500 • Fax: +44 (0)1403 276501

Approved Products	**LPC Ref. No.**
M500M monitor module	69/4
M500CH control module	69/5
M500X short circuit isolator	69/6
M501M mini monitor module	69/18

Thorn Security Limited

160 Billet Road, Walthamstow, London E17 5DR

Tel: +44 (0)181 919 4000 • Fax: +44 (0)181 919 4040

Approved Products	**LPC Ref. No.**
RM520 relay module	10/87
SM520 addressable sounder driver module	10/88
DM520 conventional detector module	10/89
LI520 line isolator module	10/90
CM520 contact monitoring module	10/101
SB520 sounder booster module	10/102

Wormald Signalco A/S, Electronics Division

Stanseveien 13, PO Box 52, Kalbakken, 0901 Oslo 9, Norway

Tel: +47 22 917600 • Fax: +47 22 917601

Approved Products	**LPC Ref. No.**
AX 87IS short circuit isolator	175c/01
AX 87AD address unit	175c/02

3 SECTION 7:
SOUNDERS

INTRODUCTION

Sounders listed in this section are tested to LPCB requirements unless otherwise indicated.

With the impending publication of EN 54 part 3, applications for approval will be accepted to this standard.

Audit Testing

Once listed, a sounder is eligible for audit testing to ensure continued compliance with the test requirements.

Note

The LPCB approves and certificates products to the requirements of product standards.

Where practical, the LPCB, in consultation with the manufacturer, will also verify that the products can meet the recommendations of applicable codes of practice.

The LPCB uses national and international standards for the listing of products. In some instances the requirements of these standards may conflict with the recommendations of local codes of practice. When this situation arises, it is on the requirements of the product standard that the LPCB base approval and certification.

If full compliance with an installation code of practice is necessary, confirmation should be obtained that the listed products also comply with the appropriate codes of practice.

Caradon Gent Limited

140 Waterside Road, Hamilton Industrial Park, Leicester LE5 1TN

Tel: +44 (0)116 2462000 • Fax: +44 (0)116 2462300

Approved Products	LPCB Ref. No.
13420-02[1] alarm sounder, 2-way	34/27
13420-03[1] alarm sounder, 3-way T-breaker	34/28
12511-37 alarm sounder	34/29
12511-52 alarm sounder	34/30
34202 alarm sounder, 2 way	042j/01
34203 alarm sounder, 3 way	042j/02
32202 alarm sounder , 2 way	042j/03
78400-02NM alarm sounder, 2 way	042j/04
32203 alarm sounder, 3 way	042j/05
78400-03NM alarm sounder, 3 way	042j/06

(1) Models manufactured with white mouldings are also Certificated; they have the same model numbers with the suffix 'WH' appended.

Middle East Sales Enquiries: MK Middle East Marketing Office, Dubai
Tel: +971 4 668 483 • Fax: +971 4 681 585

India Sales Enquiries: MK India, Madras
Tel: +91 44 626 9991 • Fax: +91 44 626 9992

Far East Sales Enquiries: MK Electric (Singapore) Pte. Limited
Tel: +65 271 7266 • Fax: +65 274 9219

The following standards are used for the evaluation of cables:

BS 6387:   *Performance requirements for cables required to maintain circuit integrity under fire condition.*

BS 6425:   *Gases evolved during combustion of electric cables. Part 1: Method for determination of amount of halogen acid gas evolved during combustion of polymeric materials taken from cables.*

BS 7629:   *Thermosetting insulated cables with limited circuit integrity when affected by fire. Part 1: Multicore cables, Part 2: Multipair cables.*

IEC 331:   *Fire resisting characteristics of electric cables.*

BS 7622: Part 2: *Measurement of smoke density of electric cables burning under defined conditions.*

Note : BS 7629 is a complete cable specification and incorporates the requirements of test specifications BS 6387 and BS 6425 : Part 1.

For cables which are considered to be acceptable for use in fire alarm installations refer to clause 17.3(b) of BS 5839 : Part 1 *Fire detection and alarm systems for buildings. Code of practice for system design, installation and servicing.*

The rated voltages recognised for BS 6387 are Uo/U, 300/500V and 450/750V, and for BS 7629 300/500V, where Uo is the power frequency voltage to earth and U is the power frequency between conductors.

Guide to Fire resistance categories from BS 6387

Resistance to fire alone
- A   650 °C for 3 hours
- B   750 °C for 3 hours
- C   950 °C for 3 hours
- S   950 °C for 20 minutes

Resistance to fire with water   -   W

Resistance to fire with mechanical shock
- X   650 °C
- Y   750 °C
- Z   950 °C

AEI Cables Limited

Birtley, Chester-le-Street, Co. Durham DH3 2RA

Tel: +44 (0)191 410 3111 • Fax: +44 (0)191 410 8312

Certificate No: 221a to BS 7629: Part 1

Firetec Multicore LSZH Cable.

LPCB Ref. No. 221a/01

Nominal csa of conductor (mm²)	Core construction (excluding drain wire and earth)	BS 6387	BS 6425 : Part 1 (outer covering)
1.0	Twin, 3, 4, 7, 12 & 19	C, W, Z	<0.5% HCl
1.5	Twin, 3, 4, 7, 12 & 19	C, W, Z	<0.5% HCl
2.5	Twin, 3, 4, 7 & 12	C, W, Z	<0.5% HCl
4.0	Twin, 3, 4 & 7	C, W, Z	<0.5% HCl

Uo/U 300/500V
Note: The Firetec Multicore cable meets the requirements of BS 7629: Part 1 which incorporates BS 6387 and BS 6425: Part1.

Certificate No: 221c to IEC 331 and BS 6425 : Part 1.

Firetec Single Core LSZH Cable.

LPCB Ref. No. 221c/01

Nominal csa of conductor (mm²)	Core construction (excluding drain wire and earth)	IEC 331 (See Notes 1 & 2)	BS 6425 : Part 1 (outer covering)
1.0	One	Complies	<0.5% HCl
1.5	One	Complies	<0.5% HCl
2.5	One	Complies	<0.5% HCl
4.0	One	Complies	<0.5% HCl
6.0	One	Complies	<0.5% HCl

Uo/U 300/500 V
Notes: (1) IEC 331 "Fire resisting characteristics of electric cables" is similar to BS 6387, Category "B".
(2) The Firetec single core LSZH cable also met IEC 331 when installed in steel conduit.

Certificate No: 221b to BS 6387, BS 6425 : Part 1 and BS 7622 : Part 2
Firetec Armoured LSZH Cable.

LPCB Ref. No. 221b/01

Nominal csa of conductor (mm²)	Core construction (excluding drain wire and earth)	BS 6387 (See Note 1)	BS 6425 : Part 1 (outer covering)	BS 7622: Part 2
1.5	Twin, 3, 4, 5, 6 and 7	C, W, Z	<0.5% HCl	>60%
2.5	Twin, 3, 4, 5, 6 and 7	C, W, Z	<0.5% HCl	>60%
4.0	Twin, 3, 4, 5, 6 and 7	C, W, Z	<0.5% HCl	>60%
6.0	Twin, 3, 4, 5	C, W, Z	<0.5% HCl	>60%
10.0	Twin, 3, 4	C, W, Z	<0.5% HCl	>60%

Uo/U 450/750V
Note 1: The Firetec armoured LSZH cable also met Category C, W, Z when tested for voltage rating 600/1000V which is not recognised by BS 6387.

Firetec Power LSZH Cable.

LPCB Ref. No. 221b/02

Nominal csa of conductor (mm²)	Core construction (excluding drain wire and earth)	BS 6387 (See Note 1)	BS 6425 : Part 1 (outer covering)	BS 7622: Part 2
1.0	Twin, 3 and 4	C, W, Z.	<0.5% HCl	>60%
1.5	Twin, 3 and 4	C, W, Z.	<0.5% HCl	>60%
2.5	Twin, 3 and 4	C, W, Z.	<0.5% HCl	>60%
4.0	Twin, 3 and 4	C, W, Z.	<0.5% HCl	>60%
6.0	Twin, 3 and 4	C, W, Z.	<0.5% HCl	>60%
10.0	Twin, 3 and 4	C, W, Z.	<0.5% HCl	>60%
16.0	Twin, 3 and 4	C, W, Z.	<0.5% HCl	>60%

Uo/U 450/750V
Note 1 : The Firetec Power LSZH cable also met Category C, W, Z when tested for voltage rating 600/1000V which is not recognised by BS 6387.

Alcatel Cable

170 Avenue Jean Jaurès, 69353 Lyon, Cedex 07, France

Tel: +33 472 72 24 20 • Fax: +33 472 72 24 02

Certificate No. 253a to BS 6387 and BS 6425 : Part 1.
Pyrolyon 'E' Fire Resistant Cable.

LPCB Ref. No. 253a/01

Nominal csa of conductor (mm²)	Core construction (excluding drain wire and earth)	BS 6387	BS 6425 : Part 1 (outer covering)
1.0	Twin, 3 & 4	C, W, Z	<0.5% HCl
1.5	Twin, 3 & 4	C, W, Z	<0.5% HCl
2.5	Twin, 3 & 4	C, W, Z	<0.5% HCl

Uo/U 300/500 V

Certificate No. 253b to BS 6387 and BS 6425 : Part 1
"X" Fire Resistant Cable.

LPCB Ref. No. 253b/01

Nominal csa of conductor (mm²)	Core construction (excluding drain wire and earth)	BS 6387	BS 6425 : Part 1 (outer covering)
1.0	Twin, 3 & 4	C, W, Z	<0.5% HCl
1.5	Twin, 3 & 4	C, W, Z	<0.5% HCl
2.5	Twin, 3 & 4	C, W, Z	<0.5% HCl

Uo/U 300/500 V

Certificate No. 253c to BS 6387 and BS 6425 : Part 1
Inferno Fire Resistant Cable.

LPCB Ref. No. 253c/01

Nominal csa of conductor (mm²)	Core construction (excluding drain wire and earth)	BS 6387	BS 6425 : Part 1 (outer covering)
1.0	Twin, 3 & 4	C, W, Z	<0.5% HCl
1.5	Twin, 3 & 4	C, W, Z	<0.5% HCl
2.5	Twin, 3 & 4	C, W, Z	<0.5% HCl

Uo/U 300/500 V

MX 331 cable

LPCB Ref. No. 253d/01

Nominal csa of Conductor (mm²)	Core construction	BS6387	BS6425 : Part 1 (See Notes 1, 2, 5	BS 7622 : Part 2 (See Note 6)	IEC 331
0.75	one[3]	C, W, Z	<0.5%HCl	>50%	Complies
1.0	one[3]	C, W, Z	<0.5%HCl	>50%	Complies
1.5	one[3]	C, W, Z	<0.5%HCl	>50%	Complies
2.5	one[3]	C, W, Z	<0.5%HCl	>50%	Complies
4.0	one[3]	C, W, Z	<0.5%HCl	>50%	Complies
6.0	one[3]	C, W, Z	<0.5%HCl	>50%	Complies
10.0	one[3]	C, W, Z	<0.5%HCl	>50%	Complies
16.0	one[3]	C, W	<0.5%HCl	>60%	Complies
25.0	one[3]	C, W	<0.5%HCl	>60%	Complies
35.0	one[4]	C, W	<0.5%HCl	>60%	Complies
50.0	one[4]	C, W	<0.5%HCl	>60%	Complies
70.0	one[4]	C, W	<0.5%HCl	>60%	Complies
95.0	one[4]	C, W	<0.5%HCl	>60%	Complies
120.0	one[4]	C, W	<0.5%HCl	>60%	Complies
150.0	one[4]	C	<0.5%HCl	>60%	Complies
185.0	one[4]	C	<0.5%HCl	>60%	Complies
240.0	one[4]	C	<0.5%HCl	>60%	Complies
300.0	one[4]	C	<0.5%HCl	>60%	Complies
400.0	one		<0.5%HCl	>60%	Complies
500.0	one		<0.5%HCl	>60%	Complies
630.0	one		<0.5%HCl	>60%	Complies

Uo/U 450/750V
Notes:
1 Clause 8 (Bending) and 9.1 & 9.2 (Impact) can not be conducted due to the cable having only one core.
2 The MX 331 cable also met category C,W,Z when tested for voltage rating 600/1000Vwhich is not recognised by BS6387.
3 To satisfy the requirement of BS 6387, testing for C, W, and Z categories was conducted using a 20mm steel conduit as the other metallic element.
4 To satisfy the requirement of BS 6387, testing for C category was conducted using a 38mm steel conduit as the other metallic element.
5 Where a single cable is fitted in a conduit, only phase to earth voltage was applied.
6 Testing to IEC 331 was conducted in conduit at 600/1000V.

3 SECTION 8:
FIRE RESISTANT CABLES

BICC Industrial Cables

Elastomeric Cables Unit, Leigh 1 Factory, Leigh, Lancashire WN7 4HB

Tel: +44 (0)1942 263004 • Fax: +44 (0)1942 263049

Certificate No: 380a to BS 6387, IEC 331 & BS 6425: Part 1

FLAMBICC Armoured   LPCB Ref. No: 380a/01

Nominal csa of conductor (mm²)	Core construction (excluding drain wire and earth)	BS 6387 (See notes 1 & 2)	BS 6425 : Part 1 (outer covering)	IEC 331 (See note 3)
1.5	Twin, 3 & 4	C, W, Z	<0.5%HCl	Complies
2.5	Twin, 3 & 4	C, W, Z	<0.5%HCl	Complies
4.0	Twin, 3 & 4	C, W, Z	<0.5%HCl	Complies
6.0	Twin, 3 & 4	C, W, Z	<0.5%HCl	Complies
10.0	Twin, 3 & 4	C, W, Z	<0.5%HCl	Complies
16.0	Twin, 3 & 4	C, W, Z	<0.5%HCl	Complies
25.0	Twin, 3 & 4	C, W, Z	<0.5%HCl	Complies
35.0	Twin, 3 & 4	C, W, Z	<0.5%HCl	Complies
50.0	Twin, 3 & 4	C, W, Z	<0.5%HCl	Complies
70.0	Twin, 3 & 4	C, W, Z	<0.5%HCl	Complies
95.0	Twin, 3 & 4	C, W, Z	<0.5%HCl	Complies
120.0	Twin, 3 & 4	C, W, Z	<0.5%HCl	Complies
150.0	Twin, 3 & 4	C, W, Z	<0.5%HCl	Complies
185.0	Twin, 3 & 4	C, W, Z	<0.5%HCl	Complies
240.0	Twin, 3 & 4	C, W, Z	<0.5%HCl	Complies
300.0	Twin, 3 & 4	C, W, Z	<0.5%HCl	Complies
400.0	Twin, 3 & 4	C, W, Z	<0.5%HCl	Complies

Uo/U 450/750V

Notes:
1) The FLAMBICC Armoured cable also met category C,W,Z when tested for voltage rating 600/1000V which is not recognised by BS6387.
2) The category Z tests were conducted in accordance with guidance from the manufacturer.
3) Testing to IEC 331 was conducted at 600/1000V.

BICC Pyrotenax

PO Box No 20, Prescot, Merseyside L3 5GB

Sales enquiries: BICC Construction Cables, PO Box 50, Helsby, Cheshire WA6 0FA

Tel: +44 (0)151 430 4050 • Fax: +44 (0)151 430 4060

Technical enquiries: Tel: +44 (0)151 430 4052 • Fax: +44 (0)151 430 4060

Certificate No:. 063a to BS 6387 and BS 6425 : Part 1

Pyrotenax Mineral Insulated Cable "Multiconductor", ZH - LSF outer covering.   LPCB Ref. No: 063a/03

Nominal csa of conductor (mm²)	Conductor construction	BS 6387	BS 6425 : Part 1 (outer covering)
1.0	Twin, 3, 4 & 7[1]	C, W, Z	<0.5% HCl
1.5	Twin, 3, 4 & 7[1]	C, W, Z	<0.5% HCl
2.5	Twin, 3, 4 & 7[1]	C, W, Z	<0.5% HCl
4.0	Twin[1]	C, W, Z	<0.5% HCl
1.0	Twin, 3, 4, 7, 12 & 19[2]	C, W, Z	<0.5% HCl
1.5	Twin, 3, 4, 7, 12 & 19[2]	C, W, Z	<0.5% HCl
2.5	Twin, 3, 4, 7 & 12[2]	C, W, Z	<0.5% HCl
4.0	Twin, 3, 4 & 7[2]	C, W, Z	<0.5% HCl
6.0	Twin, 3 & 4[2]	C, W, Z	<0.5% HCl
10.0	Twin, 3 & 4[2]	C, W, Z	<0.5% HCl
16.0	Twin, 3 & 4[2]	C, W, Z	<0.5% HCl
25.0	Twin, 3 & 4[2]	C, W, Z	<0.5% HCl

(1) Uo/U 300/500V
(2) Uo/U 450/750V

Continued

BICC Pyrotenax (continued)

Also certificated in the above construction for :
Bare (BS 6425 : Part 1 not applicable)   **LPCB Ref. No. 063a/01**
PVC outer covering not tested to comply with BS 6425 : Part 1.   **LPCB Ref. No. 063a/02**

Pyrotenax Mineral Insulated Cable "single conductor", ZH - LSF outer covering   **LPCB Ref. No. 063a/05**

Nominal csa of conductor (mm²)	Conductor construction	BS 6387	BS 6425 : Part 1 (outer covering)
10.0	One	C, W, Z	<0.5% HCl
16.0	One	C, W, Z	<0.5% HCl
25.0	One	C, W, Z	<0.5% HCl
35.0	One	C, W, Z	<0.5% HCl
50.0	One	C, W, Z	<0.5% HCl
70.0	One	C, W, Z	<0.5% HCl
95.0	One	C, W, Z	<0.5% HCl
120.0	One	C, W, Z	<0.5% HCl
150.0	One	C, W, Z	<0.5% HCl
185.0	One	C, W, Z	<0.5% HCl
240.0	One	C & W	<0.5% HCl

Uo/U 450/750V
Also certificated in the above construction for:
Bare, (BS 6425 : Part 1 not applicable)   **LPCB Ref. No. 063a/04**

Thomas Bolton Flexible Cable Limited
Waddicar Lane, Melling, Liverpool L31 1DF
Sales enquiries: BICC Construction Cables, PO Box 50, Helsby, Cheshire WA6 0FA

Tel: +44 (0)1928 727 2727• Fax: +44 (0)1928 762676
Technical enquiries: Tel: +44 (0)1928 762601 • Fax: +44 (0)1928 762531 **LPCB**

Certificate No. 317a to BS 6387 and BS 6425 : Part 1
Flamsil FRC   **LPCB Ref. No. 317a/01**

Nominal csa of conductor (mm²)	Core construction (excluding drain wire and earth)	BS 6387	BS 6425 : Part 1 (outer covering)
1.0	Twin, 3 and 4	C, W, Z	<0.5% HCl
1.5	Twin, 3 and 4	C, W, Z	<0.5% HCl
2.5	Twin, 3 and 4	C, W, Z	<0.5% HCl
4.0	Twin, 3 and 4	C, W, Z	<0.5% HCl

Uo/U 300/500V

Dätwyler AG
CH-6460 Altdorf, Switzerland

Tel: +41 875 1122 • Fax: +41 875 1870
UK Sales Enquiries: Tel: +44 (0)1462 482888 • Fax: +44 (0)1462 481038 **LPCB**
Certificate No: 172a to BS 6387 and BS 6425 : Part 1

Lifeline cables, low smoke zero halogen outer covering.   **LPCB Ref. No. 172a/01**

Nominal csa of conductor (mm²)	Core construction (excluding drain wire and earth)	BS 6387	BS 6425 : Part 1 (outer covering)
1.0	Twin, 3 and 4	C, W, Z	<0.5% HCl
1.5	Twin, 3 and 4	C, W, Z	<0.5% HCl
2.5	Twin, 3 and 4	C, W, Z	<0.5% HCl

Uo/U 300/500V

3 SECTION 8:
FIRE RESISTANT CABLES

Delta Special Cables Limited
Manston Lane, Crossgates, Leeds LS15 8SZ
Sales Enquiries: Tel: +44 (0)645 318300 • Fax: +44 (0)113 232 1633

Certificate No: 076a to BS 6387 and BS 6425: Part 1
076b to BS 7629

Firetuf OHLS Multipair.   LPCB Ref. No. 076a/01

Nominal csa of conductor (mm)	Core construction (excluding drain wire and earth)	BS 6387	BS 6425 : Part 1 (outer covering)
1.0	1, 2, 5 and 10 pairs	C, W, Z	<0.5% HCl
1.5	1, 2 and 5 pairs	C, W, Z	<0.5% HCl
2.5	1 and 2 pairs	C, W, Z	<0.5% HCl
1.0	15 and 20 pairs	C & W	<0.5% HCl
1.5	10, 15 and 20 pairs	C & W	<0.5% HCl
2.5	5, 10, 15 and 20 pairs	C & W	<0.5% HCl

Uo/U 300/500 V

Firetuf OHLS Multicore.   LPCB Ref. No. 076a/02

Nominal csa of conductor (mm^2)	Core construction (excluding drain wire and earth)	BS 6387	BS 6425 : Part 1 (outer covering)
1.0	12 & 19	C, W, Z	<0.5% HCl

Uo/U 300/500 V

Firetuf OHLS Multicore.   LPCB Ref. No. 076b/01

Nominal csa of conductor (mm^2)	Core construction (excluding drain wire and earth)	BS 6387	BS 6425 : Part 1 (outer covering binder tape, Insulation)
1.0	Twin, 3, 4, & 7	C, W, Z	<0.5% HCl
1.5	Twin, 3, 4, 7, 12 & 19	C, W, Z	<0.5% HCl
2.5	Twin, 3, 4, 7, & 12	C, W, Z	<0.5% HCl
4.0	Twin, 3, 4	C, W, Z,	<0.5%HCl

Uo/U 300/500V

Note: Firetuf OHLS Multicore meets the requirements of BS 7629 which incorporates the requirements of BS 6387 and BS 6425 : Part 1.

Delta Energy Cables Limited
Alfreton Road, Derby DE21 4AE

Tel: +44 (0)1332 345431 • Fax: +44 (0)1332 376373

Certificate No: 361a to IEC 331 and BS 6425: Part 1.
Sifer 950i Single core fire resistant OHLS Cable: **LPCB Ref No: 361a/01**

Nominal csa of conductor (mm²)	Core Construction	IEC 331 (see notes 1, 2 and 3).	BS 6425: Part 1 (outer covering)
1.5	One	Complies	<0.5% HCl
2.5	One	Complies	<0.5% HCl
4.0	One	Complies	<0.5% HCl
6.0	One	Complies	<0.5% HCl
10.0	One	Complies	<0.5% HCl
16.0	One	Complies	<0.5% HCl
25.0	One	Complies	<0.5% HCl
35.0	One	Complies	<0.5% HCl
50.0	One	Complies	<0.5% HCl
70.0	One	Complies	<0.5% HCl
95.0	One	Complies	<0.5% HCl
120.0	One	Complies	<0.5% HCl
150.0	One	Complies	<0.5% HCl
185.0	One	Complies	<0.5% HCl
240.0	One	Complies	<0.5% HCl
300.0	One	Complies	<0.5% HCl
400.0	One	Complies	<0.5% HCl
500.0	One	Complies	<0.5% HCl
630.0	One	Complies	<0.5% HCl

Uo/U 600/1000V

Sifer 950s Single core fire resistant OHLS Cable: **LPCB Ref No: 361a/02**

Nominal csa of conductor (mm²)	Core Construction	IEC 331 (see notes 1, 2 and 3).	BS 6425: Part 1 (outer covering)
1.5	One	Complies	<0.5% HCl
2.5	One	Complies	<0.5% HCl
4.0	One	Complies	<0.5% HCl
6.0	One	Complies	<0.5% HCl
10.0	One	Complies	<0.5% HCl
16.0	One	Complies	<0.5% HCl
25.0	One	Complies	<0.5% HCl
35.0	One	Complies	<0.5% HCl
50.0	One	Complies	<0.5% HCl
70.0	One	Complies	<0.5% HCl
95.0	One	Complies	<0.5% HCl
120.0	One	Complies	<0.5% HCl
150.0	One	Complies	<0.5% HCl
185.0	One	Complies	<0.5% HCl
240.0	One	Complies	<0.5% HCl
300.0	One	Complies	<0.5% HCl
400.0	One	Complies	<0.5% HCl
500.0	One	Complies	<0.5% HCl
630.0	One	Complies	<0.5% HCl

Uo/U 600/1000V

Notes: 1. IEC 331 Fire resisting characteristics of electric cables is similar to BS 6387 Category B.
         2. The Sifer 950i and 950s cable also met IEC 331: when tested at 950°c which is similar to BS6387 category C.
         3 The Sifer 950i and 950s cables (up to and including 25mm²) also met IEC 331 when tested in conduit at 750°C and 950°C which is not recognised by the standard.

Continued

3 SECTION 8:
FIRE RESISTANT CABLES

Delta Energy Cables Limited (continued)

Certificate No: 361b to BS 6387 and BS 6425 : Part 1
Firetuf Power OHLS cable

LPCB Ref. No. 361b/01

Nominal csa of conductor (mm²)	Core construction (excluding drain wire and earth)	BS 6387 (See notes 1 & 2)	BS 6425 : Part 1 (outer covering)
1.5	Twin 3 & 4	C.W.Z.	<0.5% HC1
2.5	Twin 3 & 4	C.W.Z.	<0.5% HC1
4.0	Twin 3 & 4	C.W.Z.	<0.5% HC1
6.0	Twin 3 & 4	C.W.Z.	<0.5% HC1
10.0	Twin 3 & 4	C.W.Z.	<0.5% HC1
16.0	Twin 3 & 4	C.W.Z.	<0.5% HC1
25.0	Twin 3 & 4	C.W.Z.	<0.5% HC1
35.0	Twin 3 & 4	C.W.Z.	<0.5% HC1
50.0	Twin 3 & 4	C.W.Z.	<0.5% HC1
70.0	Twin 3 & 4	C.W.Z.	<0.5% HC1
95.0	Twin 3 & 4	C.W.Z.	<0.5% HC1
120.0	Twin 3 & 4	C.W.Z.	<0.5% HC1
150.0	Twin 3 & 4	C.W.Z.	<0.5% HC1
185.0	Twin 3 & 4	C.W.Z.	<0.5% HC1
240.0	Twin 3 & 4	C.W.Z.	<0.5% HC1
300.0	Twin 3 & 4	C.W.Z.	<0.5% HC1

Uo/U 450/750V

Notes: 1 The Firetuf Power OHLS cable also met category C.W.Z. when tested for voltage rating 600/1000V which is not recognised by BS6387.
2 For cables with an outside diameter in excess of 20mm, the Category Z tests were conducted in accordance with guidance from the manufacturer.

Certificate No: 361c to BS 6387 and BS 6425: Part 1
Firetuf® FTPU 950 fire resistant Mifer OHLS® cable

LPCB Ref No: 361c/02

Nominal csa of conductor (mm²)	Core Construction (excluding drain wire and earth)	BS 6387 (see notes 1 & 2).	BS 6425: Part 1 (outer covering)
1.5	Twin, 3, 4 & 5	C,W,Z	<0.5%HCl
2.5	Twin, 3, 4 & 5	C,W,Z	<0.5%HCl
4.0	Twin, 3, 4 & 5	C,W,Z	<0.5%HCl
6.0	Twin, 3, 4 & 5	C,W,Z	<0.5%HCl
10.0	Twin, 3, 4 & 5	C,W,Z	<0.5%HCl
16.0	Twin, 3, 4 & 5	C,W,Z	<0.5%HCl
25.0	Twin, 3, 4 & 5	C,W,Z	<0.5%HCl
35.0	Twin, 3, 4 & 5	C,W,Z	<0.5%HCl
50.0	Twin, 3, 4 & 5	C,W,Z	<0.5%HCl
70.0	Twin, 3, 4 & 5	C,W,Z	<0.5%HCl
95.0	Twin, 3, 4 & 5	C,W,Z	<0.5%HCl
120.0	Twin, 3 & 4	C,W,Z	<0.5%HCl
150.0	Twin, 3 & 4	C,W,Z	<0.5%HCl
185.0	Twin, 3 & 4	C,W,Z	<0.5%HCl
240.0	Twin, 3 & 4	C,W,Z	<0.5%HCl
300.0	Twin, 3 & 4	C,W,Z	<0.5%HCl
4000	Twin, 3 & 4	C,W,Z	<0.5%HCl
50/25	4+ Concentric Earth	C,W,Z	<0.5%HCl
70/35	4+ Concentric Earth	C,W,Z	<0.5%HCl
95/50	4+ Concentric Earth	C,W,Z	<0.5%HCl
120/70	4+ Concentric Earth	C,W,Z	<0.5%HCl
150/70	4+ Concentric Earth	C,W,Z	<0.5%HCl
185/95	4+ Concentric Earth	C,W,Z	<0.5%HCl
240/120	4+ Concentric Earth	C,W,Z	<0.5%HCl
Multicore			
1.5	7,10,12,16,20,21 & 30	C,W,Z	<0.5%HCl
2.5	7,10,12,16,20,21 & 30	C,W,Z	<0.5%HCl

Uo/U 450/750V

Notes:
1 The Firetuf® FTPU 950 fire resistant Mifer OHLS® cable also met category C,W,Z when tested for voltage rating 600/1000V which is not recognised by BS 6387.
2 For cable with an outside diameter of in excess of 20mm, the category Z tests were conducted in accordance with guidance from the manufacturer.

Delta Energy Cables Limited

Copperworks Road, Llanelli, Dyfed SA15 2NH

Tel: +44 (0)1554 750121 • Fax: +44 (0)1554 775502

LPCB

Certificate No: 359a to IEC 331 and 6425: Part 1.
Sifer 950i Single core fire resistant OHLS cable.   LPCB Ref No: 359a/01

Nominal csa of conductor (mm²)	Core Construction	IEC 331 (see notes 1, 2 and 3).	BS 6425: Part 1 (outer covering)
1.5	One	Complies	<0.5% HCI
2.5	One	Complies	<0.5% HCI
4.0	One	Complies	<0.5% HCI
6.0	One	Complies	<0.5% HCI
10.0	One	Complies	<0.5% HCI
16.0	One	Complies	<0.5% HCI

Uo/U 600/1000V

Sifer 950s Single core fire resistant OHLS cable.   LPCB Ref No: 359a/02

Nominal csa of conductor (mm²)	Core Construction	IEC 331 (see notes 1, 2 and 3).	BS 6425: Part 1 (outer covering)
1.5	One	Complies	<0.5% HCI
2.5	One	Complies	<0.5% HCI
4.0	One	Complies	<0.5% HCI
6.0	One	Complies	<0.5% HCI
10.0	One	Complies	<0.5% HCI
16.0	One	Complies	<0.5% HCI

Uo/U 600/1000V
Notes: 1. IEC 331 Fire resisting characteristics of electric cables is similar to BS 6387 Category B.
2. The Sifer 950i and of 950s cables also met IEC 331: when tested at 950°c which is similar to BS 6387, category C.
3. The Sifer 950i and 950s cables also met IEC 331: when tested in conduit at 750°C and 950°C which is not recognised by the standard

Doncaster Cables

Millfields Industrial Estate, Arksey Lane, Bentley, Doncaster, South Yorkshire DN5 0SJ

Tel: +44 (0)1302 873408 • Fax: +44 (0)1302 820425

LPCB

Certificate No. 338a to BS 7629: Part 1 and BS 7622: Part 2

Firesure L.D.   LPCB Ref.No. 338a/01

Nominal csa of conductor (mm2)	Core construction	BS 6387 (see notes)	BS6425 : Part 1 (outer covering)	BS 7622: Part 2
1.0	Twin, 3, 4	C,W,Z	<0.5% HC1	Complies
1.5	Twin, 3, 4	C,W,Z	<0.5% HC1	Complies
2.5	Twin, 3, 4	C,W,Z	<0.5% HC1	Complies
4.0	Twin, 3, 4	C,W,Z	<0.5% HC1	Complies

Uo/U 300/500V

Note 1: Firesure LD meets the requirements of BS 7629 which incorporates the requirements of BS6387 and BS 6425 : Part 1

3 SECTION 8:
FIRE RESISTANT CABLES

Fabbrica Conduttori Electtrici Cavicel SpA,
Via G. D'Annunzio, 44-20096 Pioltello, Milan, Italy

Tel: +39 2 921 60521 • Fax: +39 2 921 60753

Certificate Nos: 217c to BS 6387 and BS 6425 : Part 1.

Firecel SR/114 H Fire Resistant Cable

LPCB Ref. No. 217c/01

Nominal csa of conductor (mm²)	Core construction (excluding drain wire and earth)	BS 6387	BS 6425 : Part 1 (outer covering)
1.0	Twin, 3 and 4	C, W, Z	< 0.5% HCl
1.5	Twin, 3 and 4	C, W, Z	< 0.5% HCl
2.5	Twin, 3 and 4	C, W, Z	< 0.5% HCl
4.0	Twin, 3 and 4	C, W, Z	<0.5% HCl

Uo/U 300/500 V

Huber & Suhner AG
Tumbelenstrasse 20, CH-8330 Pfäffikon/ZH, Switzerland

Tel: +41 1 952 22 11 • Fax: +41 1 952 24 24
UK Sales Enquiries: Suhner Electronics Limited.
Tel: +44 (0)1869 244676 • Fax: +44 (0)1869 249046

Certificate No: 075a to BS 6387 and BS 6425 : Part 1

Radox FR control and power cables, low smoke zero halogen outer covering

LPCB Ref. No. 075a/01

Nominal csa of conductor (mm²)	Core construction (excluding drain wire and earth)	BS 6387	BS 6425 : Part 1 (outer covering)
0.75	Twin, 3, 4, 5, 6 and 7[1]	C, W, Z	<0.5% HCl
1.0	Twin, 3, 4, 5, 6 and 7[1]	C, W, Z	<0.5% HCl
1.5	Twin, 3, 4, 5, 6, 7, 10 and 16[2]	C, W, Z	<0.5% HCl
2.5	Twin, 3, 4, 5, 6, 7 and 10[2]	C, W, Z	<0.5% HCl
4.0	Twin, 3, 4, 5, 6 and 7[2]	C, W, Z	<0.5% HCl
6.0	Twin, 3, 4, 5, 6 and 7[2]	C, W, Z	<0.5% HCl

(1) Uo/U 300/500V (2) Uo/U 450/750V

Radox FR Communication Cable, low smoke zero halogen outer covering.

LPCB Ref. No. 075a/02

Nominal csa of conductor (mm²)	Core construction (excluding drain wire and earth)	BS 6387	BS 6425 : Part 1 (outer covering)
1.5mm²	1 pair	C, W, Z	<0.5% HCl

Uo/U 300/500V

Irish Driver-Harris Company Limited
Kilkenny Industrial & Business Park, Purellsinch, Kilkenny, Ireland

Tel: +353 56 70800 • Fax: +353 56 80808

Certificate No. 408a to BS 7629 : Part 1

Kilflam 1082

LPCB Ref. No. 408a/01

Nominal csa of conductor (mm²)	Core construction	BS 6387	BS 6425 : Part 1 Outer covering, insulation and binder tape
1.0	Twin 3 & 4	C,W,Z	<0.5%HC1
1.5	Twin 3 & 4	C,W,Z	<0.5%HC1
2.5	Twin 3 & 4	C,W,Z	<0.5%HC1
4.0	Twin 3 & 4	C,W,Z	<0.5%HC1

Pirelli Cables Limited

Energy Cables, Chickenhall Lane, Bishopstoke, Eastleigh, Hants SO50 6YU

Tel: +44 (0)990 133143 • Fax: +44 (0)1703 295 151

FP 100 fire resistant single core cable for installation in conduit   LPCB Ref. No. 077c/01

Nominal csa of conductor (mm²)	Core Construction	BS 6387 (See Notes)	BS 6425 : Part 1 (outer covering)
1.0	One	C, W, Z	<0.5% HCl
1.5	One	C, W, Z	<0.5% HCl
2.5	One	C, W, Z	<0.5% HCl
4.0	One	C, W, Z	<0.5% HCl
6.0	One	C, W, Z	<0.5% HCl
10.0	One	C, W, Z	<0.5% HCl

Uo/U 450/750 V

Notes: (1) Testing was conducted using 20mm steel conduit.
       (2) To satisfy the requirement of BS 6387 the conduit was considered as the other metallic element.
       (3) Clauses 8 (Bending) and 9.1 & 9.2 (Impact) were not conducted.

Certificate No. 077d to BS 6387, BS 6425: Part 1 and BS 7622: Part 2
FP150 LS0H   LPCB Ref. No. 077d/01

Nominal csa of Conductor (mm²)	Core construction (excluding drain wire and earth)	BS6387	BS6425: Part 1 (outer covering)	BS7622: Part 2
1.0	Twin, 3 & 4	C,W,Z	<0.5% HCl	Complies
1.5	Twin, 3 & 4	C,W,Z	<0.5% HCl	Complies
2.5	Twin, 3 & 4	C,W,Z	<0.5% HCl	Complies

Uo/U 300/500 V

Certificate No: 077a to BS 7629
         077b to BS 6387 and BS 6425: Part 1

FP 200 LSOH   LPCB Ref. No. 077a/01

Nominal csa of conductor (mm²)	Core construction (excluding drain wire and earth)	BS 6387 (See note 1)	BS 6425 : Part 1 (outer covering binder tape)
1.0	Twin, 3 & 4	C, W, Z	<0.5% HCl
1.5	Twin, 3 & 4	C, W, Z	<0.5% HCl
2.5	Twin, 3 & 4	C, W, Z	<0.5% HCl
4.0	Twin, 3 & 4	C, W, Z	<0.5% HCl

Uo/U 300/500V

Note - 1 FP 200 LSOH meets the requirements of BS 7629 which incorporates the requirements of BS 6387 and BS 6425: Part 1.

FP 200 GOLD LSOH.   LPCB Ref No. 077a/02

Nominal csa of conductor (mm²)	Core Construction (excluding drain wire and earth)	BS 6387 (See Note 1)	BS 6425 : Part 1 (outer covering, binder tape)
1.0	Twin, 3, 4, & 7	C, W, Z	<0.5%HCl
1.5	Twin, 3, 4, 7, 12 & 19	C, W, Z	<0.5%HCl
2.5	Twin, 3, 4, 7 & 12	C, W, Z	<0.5%HCl
4.0	Twin, 3 & 4	C, W, Z	<0.5%HCl

Uo/U 300/500 V

Note :FP 200 Gold LSOH meets the requirements of BS 7629 which incorporates the requirements of BS 6387 and BS 6425 : Part 1.

Continued

Pierlli Cable Limited (continued)

Certificate No. 077a to BS 7629 : Part 1
FP200 FLEX

077a/03

Nominal csa of conductor (mm²)	Core Construction (excluding drain wire and earth)	BS 6387 (See Note 1)	BS 6425 : Part 1 (outer covering, binder tape)
1.0	Twin, 3 & 4	C,W,Z	<0.5% HCl
1.5	Twin, 3 & 4	C,W,Z	<0.5% HCl
2.5	Twin, 3 & 4	C,W,Z	<0.5% HCl

Uo/U 300/500

Note: FP200 Flex meets the requirements of BS 7629 which incorporates the requirements of BS 6387 and BS 6425: Part 1.

Certificate Number 077b to BS 6387: 1994 and BS 6425: Part 1 : 1990
FP 200 GOLD LSOH.

LPCB Ref No. 077b/03

Nominal csa of conductor (mm²)	Core Construction (excluding drain wire and earth)	BS 6387	BS 6425 : Part 1 (outer covering, binder tape)
2.5	19	C & W	<0.5%HCl

Uo/U 300/500 V

FP 300 Fire resistant cable LSOH outer covering.

LPCB Ref. No. 077b/01

Nominal csa of conductor (mm²)	Core construction	BS 6387 (see Note 1)	BS 6425 : Part 1 (outer covering)
1.0	Twin, 3 and 4	C, W, Z	< 0.5% HCl
1.5	Twin, 3 and 4	C, W, Z	< 0.5% HCl
2.5	Twin, 3 and 4	C, W, Z	< 0.5% HCl
4.0	Twin, 3 and 4	C, W, Z	< 0.5% HCl
6.0	Twin, 3 and 4	C, W, Z	< 0.5% HCl
10.0	Twin, 3 and 4	C, W, Z	< 0.5% HCl
16.0	Twin, 3 and 4	C, W, Z	< 0.5% HCl

Uo/U 450/750 V

Note 1:The FP 300 LSOH cable also met Category C,W,Z when tested for voltage rating 600/1000V which is not recognised by BS 6387.

LPCB

FP 400 Fire resistant cable LSOH outer covering.

LPCB Ref. No. 077b/02

Nominal csa of conductor (mm²)	Core construction	BS 6387 (see Note 1)	BS 6425 : Part 1 (outer covering)
1.0	Twin, 3 and 4	C, W, Z	< 0.5% HCl
1.5	Twin, 3 and 4	C, W, Z	< 0.5% HCl
2.5	Twin, 3 and 4	C, W, Z	< 0.5% HCl
4.0	Twin, 3 and 4	C, W, Z	< 0.5% HCl
6.0	Twin, 3 and 4	C, W, Z	< 0.5% HCl
10.0	Twin	C, W, Z	< 0.5% HCl

Uo/U 450/750 V

Note 1:   The FP 400 LSOH cable also met Category C,W,Z when tested for voltage rating 600/1000V which is not recognised by BS 6387.

RAYDEX/CDT Limited

P O Box 3, Church Street, Littleborough, Lancashire OL15 8HG

Tel: +44 (0)1706 374015 • Fax: +44 (0)1706 370576

Certificate No. 288a to BS 6387 and BS 6425: Part 1

LPCB Ref. No. 288a/01

FG 950 Fire performance cable low smoke zero halogen outer covering

Nominal csa of conductor (mm²)	Core construction	BS 6387	BS 6425 : Part 1 (outer covering)
1.0	Twin, 3 and 4	C, W, Z	< 0.5% HCl
1.5	Twin, 3 and 4	C, W, Z	< 0.5% HCl
2.5	Twin, 3 and 4	C, W, Z	< 0.5% HCl
4.0	Twin, 3 and 4	C, W, Z	< 0.5% HCl

Uo/U 300/500 V

Tratos Cavi SpA

52036 Pieve S. Stefano (AR), Via Stadio 2, Arezzo, Italy

Tel : +39 57 57 94211 • Fax : +39 57 57 98026

Certificate No: 222a to BS 7629: 1993

FIRE- safe TW950 Fire Resistant Cable.

LPCB Ref. No. 222a/01

Nominal csa of conductor (mm²)	Core construction (excluding drain wire and earth)	BS 6387	BS 6425 : Part 1 (outer covering, insulation and binder tape)
1.0	Twin, 3 and 4	C, W, Z	<0.5% HCl
1.5	Twin, 3 and 4	C, W, Z	<0.5% HCl
2.5	Twin, 3 and 4	C, W, Z	<0.5% HCl
4.0	Twin, 3 and 4	C, W, Z	<0.5% HCl

Uo/U 300/500 V.

Note: FIRE-safe TW950 Fire Resistant Cable meets the requirements of BS 7629 which incorporates the requirements of BS 6387 and BS 6425 : Part 1.

Wrexham Mineral Cables

Wynnstay Technology Park, Ruabon, Wrexham, Clwyd LL14 6EN

Tel: +44 (0)1978 810789 • Fax: +44 (0)1978 821502

Certificate No. 333a to BS 6387 and BS 6425 : Part 1

LPCB Ref. No. 333a/01

Wrexham Mineral Insulated Cable

Nominal csa of conductor (mm²)	Core construction	BS 6387	BS 6425 : Part 1 (outer covering)
1.0[1]	Twin,3 & 4	C, W, Z	<0.5%HCl
1.5[1]	Twin,3 & 4	C, W, Z	<0.5%HCl
2.5[1]	Twin,3 & 4	C, W, Z	<0.5%HCl
4.0[1]	Twin	C, W, Z	<0.5%HCl
1.5[2]	Twin,3 & 4	C, W, Z	<0.5%HCl
2.5[2]	Twin,3 & 4	C, W, Z	<0.5%HCl
4.0[2]	Twin,3 & 4	C, W, Z	<0.5%HCl
6.0[2]	Twin,3 & 4	C, W, Z	<0.5%HCl
10.0[2]	Twin	C, W, Z	<0.5%HCl

(1)   Uo/U 300/500V
(2)   Uo/U 450/750V

3 SECTION 9:
UL 910 CABLES

INTRODUCTION

UL 910 'Test for flame propagation and smoke density values for electrical and optical fibre cables used in spaces transporting environmental air'.

The above certification scheme is now operational and LPCB are accepting applications for testing and certification.

The equipment used to conduct this testing is known as the Steiner Tunnel which has been installed at the LPC Laboratories in Borehamwood.

LPCB is conducting testing and certification in partnership with Underwriters' Laboratories, Northbrook.

Each applicant firm is also required to attain LPCB certification to ISO 9001 or ISO 9002 - 'Quality Systems' which is maintained through a regular surveillance programme.

RAYDEX/CDT Limited

P O Box 3, Church Street, Littleborough, Lancashire OL15 8HG

Tel: +44 (0)1706 374015 • Fax: +44 (0)1706 370576

Certificate No: 288b to UL910

RAYDEX Plenum Cat 5 Cable

Ref: CDF5-UTP-PLEN

Nominal diameter of conductor	Core construction	UL910	BS 6425 : Part 1 (outer covering)	BS 7622: Part 2
0.52mm/24 AWG	4 pairs	Complies	<0.5%HCl	>50%

Certificate No. 288c to UL 910
Raydex Plenum Cat 5 Cable Ref: CDP5-UTP-PLEN

Nominal diameter of conductor	Core construction	UL 910
0.52 mm/24 AWG	4 pairs	Complies

Lucent Technologies Inc

12000 I Street, Omaha, Nebraska, NE 69137-9000, USA

Tel: +1 402 691 3221 • Fax: +1 402 691 3669

Certificate No. 417a to UL 910

LUCENT Plenum Cat 5 Cable

Ref: 2061B Systimax

LPCB Ref. No. 417a/01

Nominal Diameter of Conductor	Core Construction	UL 910
0.52mm/24 AWG	4 pairs	Complies

Ref: 2071A GigaSpeed

LPCB Ref. No. 417a/02

Nominal Diameter of Conductor	Core Construction	UL 910
0.52mm/24 AWG	4 pairs	Complies

INTRODUCTION

With the introduction of BS5979 : 1993 the title of this section has been changed from Central Stations for fire alarm systems to alarm receiving centres (ARC's).

The approvals to LPS 1020 will be phased out as the listed ARC's obtain compliance with BS5979 : 1993. Approvals to LPS 1020 will be withdrawn at the end of the year 2000.

For clarity the following marks are used in the list:

LPS 1020 - Approval   BS5979 - Certification
                                         and LPS 1020 Approval

It is a requirement of LPS 1020 for the central stations to offer an LPCB Approved system for the signalling of alarms from the protected premises to the central station. This is not a requirement of BS5979 : 1993. Central stations also offer other methods of signalling which may not be approved. The scheme is operated jointly between LPCB and NACOSS.

ADT Fire and Security plc

Regency Court, 2A High Street, Kings Heath, Birmingham B14 7SW

Tel: +44 (0)121 443 1300 • Fax: +44 (0)121 444 0417

CSFA016

Alarm Receiving equipment: Telecom Red Caresystem/Industry Standard Receivers/CCTV monitoring
Approved fire authority areas of operation.
Avon, Cambridgeshire, Clwyd, Cornwall, Derbyshire, Devon, Dorset, Dyfed, Gloucestershire, Gwent, Hereford & Worcester, Leicestershire, Lincolnshire, Mid Glamorgan, Norfolk, Northamptonshire, Nottinghamshire, Oxfordshire, Powys, Shropshire, Somerset, South Glamorgan, Staffordshire, Suffolk, Warwickshire, West Glamorgan, West Midlands, Wiltshire.

Banham Alarms
(Banham Patent Locks Limited)

10 Pascal Street, London, SW8 4SH

Tel: +44 (0)171 622 5151 • Fax: +44 (0)171 498 2461

CSFA006

Alarm receiving equipment: Telecom Red CARE System.
Approved Fire Authority areas of operation:
Berkshire, Hampshire, Kent, Greater London, Surrey.

MetropolitanSecurity Services (Prop: Wigan Matropolitan Council)

Municipal Buildings, Hewlett Street, Wigan WN1 1NQ

Tel: +44 (0)1942 827445 • Fax: +44 (0)1942 827033

CSFA015

Alarm receiving equipment: Telecom Red CARE System
Approved fire authority areas of operation:
Greater Manchester.

3 SECTION 10:
ALARM RECEIVING CENTRERS

National Monitoring (Prop. AVR Group Limited)
Units 16/24, Attenburys Park Estate, Attenburys Lane, Timperley, Cheshire WA14 5QN

Tel: +44 (0)161 976 4747 • Fax: +44 (0)161 973 7879

LPCB
CSFA008

Alarm receiving equipment: Telecom Red CARE System.
Approved fire authority areas of operation:
Avon, Bedfordshire, Berkshire, Cambridgeshire, Cheshire, Clwyd, Cumbria, Derbyshire, Devon, Dorset, Essex, Gloucestershire, Hampshire, Hereford & Worcester, Humberside, Lancashire, Leicestershire, Lincolnshire, London, Greater Manchester, Merseyside, Norfolk, Northamptonshire, Northern Ireland, Nottinghamshire, North Yorkshire, Shropshire, South Yorkshire, Staffordshire, Suffolk, Surrey, Tyne and Wear Warwickshire, West Glamorgan, West Midlands, West Yorkshire, Strathclyde.

Regal Security Systems Limited*
Kingston Business Centre, Fullers Way South, Chessington, Surrey KT9 1HN

Tel: +44 (0)181 397 0074 • Fax: +44 (0)181 391 2704

LPCB
CSFA009

Alarm receiving equipment: Telecom Red CARE System.
Approved fire authority areas of operation:
Berkshire, Buckinghamshire, Cheshire, Cumbria, Essex, Greater London, Greater Manchester, Hertfordshire, Kent, Lancashire, Merseyside, Surrey.

Securicor Cash Services Limited
Valiant House, 67-71 Shoreditch High Street, Shoreditch, London E1 6JJ

Tel: +44 (0)990 774411 • Fax: +44 (0)990 143055

LPCB
CSFA001

Alarm receiving equipment: Telecom Red CARE System.
Approved fire authority areas of operation:
Avon, Bedfordshire, Berkshire, Buckinghamshire, Cornwall, Devon, Dorset, East Sussex, Essex, Mid Glamorgan, Gloucestershire, Greater London, Gwent, Hampshire, Hereford & Worcester, Hertfordshire, Kent, Oxfordshire, Somerset, South Glamorgan, Surrey, West Glamorgan, West Sussex, Wiltshire.

Security Services (Prop: Bryan Enterprises Ltd.)
2 Berry Street, Stoke on Trent, ST4 1AY

Tel: +44 (0)1782 416308 • Fax: +44 (0)1782 849522

CSFA007

Alarm receiving equipment: Telecom Red CARE System.
Approved fire authority areas of operation:
Cheshire, Clwyd, Greater Manchester, Gwynedd, Shropshire, Staffordshire, West Midlands

Southern Monitoring Services Limited
Security House, 212-218, London Road, Waterlooville, Hampshire PO7 7AJ

Tel: +44 (0)1705 265113 • Fax: +44 (0)1705 256001

CSFA003

Alarm receiving equipment: Telecom Red CARE System.
Approved fire authority areas of operation:
Avon, Bedfordshire, Berkshire, Buckinghamshire, Cambridgeshire, Cheshire, Cleveland, Clwyd, Cornwall, Cumbria, Derbyshire, Devon, Dorset, East Sussex, Essex, Gloucestershire, Grampian, Greater London, Gwent, Hampshire, Hereford & Worcester, Hertfordshire, Humberside, Isle of Wight, Kent, Lancashire, Leicestershire, Lincolnshire, Lothian, Greater Manchester, Merseyside, Mid Glamorgan, Norfolk, Northamptonshire, North Yorkshire, Nottinghamshire, Oxfordshire, Powys, Shropshire, Somerset, South Glamorgan, South Yorkshire, Staffordshire, Strathclyde, Suffolk, Surrey, Tyne & Wear, Warwickshire, West Glamorgan, West Midlands, West Sussex, West Yorkshire, Wiltshire.

INTRODUCTION

The following systems are those approved by the LPCB for signalling of alarms from protected premises to central stations. It is a requirement for approval for a central station to be able to offer an LPCB approved method of signalling.

ADT Security Systems*

Titan House, 184/192 Bermondsey Street, London SE1 3UG

Tel: +44 (0)171 407 9741 • Fax: +44 (0)171 403 0907

Approved System:
1077 Multiplex Signalling System

* This company has not been assessed by LPCB to ISO 9000.

British Telecom - Telecom Red*

Proctor House, 100-110 High Holborn, London WC1V 6LD

Tel: +44 (0)171 728 8513 • Fax: +44 (0)171 728 8260

Approved System:
Telecom Red CARE Signalling System

The system components are manufactured by;

Versus Technology Limited
Unit B7, Armstrong Mall, Southwood Summit Centre, Farnborough, Hampshire GU14 0HR

Tel: +44 (0)1252 378822 • Fax: +44 (0)1252 371166

and include;

	LPCB Ref.No.
Host computer	099a/01
Scanner	099a/02
Microscanner	099a/04
EUROSTU (Subscriber Terminal Unit)	099a/06
3GSTU-12V (Subscriber Terminal Unit)	099a/07
3GSTU-24V (Subscriber Terminal Unit)	099a/08
3GSTU-PLI (Subscriber Terminal Unit)	099a/09

* This company has not been assessed by LPCB to ISO 9000.

Thorn Security Limited

160 Billet Road, Walthamstow, London E17 5DR

Tel: +44 (0)181 919 4000 • Fax: +44 (0)181 919 4040

Approved System:
Thorn Security Mainline and SE100 Systems

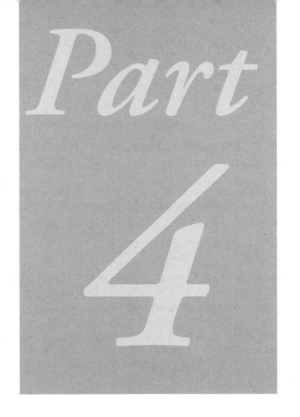

Part

4

*PORTABLE
FIRE
EXTINGUISHERS
AND FIRE
HOSE REELS*

4 PORTABLE FIRE EXTINGUISHERS AND FIRE HOSE REELS

INTRODUCTION

This part of the List covers portable fire extinguishers and fire hose reels.

IDENTIFICATION OF APPROVED EXTINGUISHERS AND FIRE HOSE REELS

All extinguishers and hose reels carry the appropriate LPCB reference number and the LPCB Mark.

Portable fire extinguishers and fire hose reels certified under the British Approvals for Fire Equipment (BAFE) scheme are identified by an asterisk (*).

TEST STANDARD

The test standard for portable extinguishers is EN 3. The test standards for hose reels are EN 671 Part 1 and EN 671 Part 2.

INSTALLATION AND MAINTENANCE OF PORTABLE FIRE EXTINGUISHING APPLIANCES

It is important that portable fire extinguishers are selected, sited, serviced and maintained in accordance with BS 5306: Part 3 or equivalent locally applied specification.

Portable fire extinguisher service and maintenance companies should also be in a position to provide certificates to this effect and to supply specifications and drawings indicating suggested sizes and types of extinguishers suitable for the hazards involved and their location.

These certificates, specifications and plans may be accepted by the fire insurers as evidence that the appliances have been properly installed and are being properly maintained.

Section 2 details LPCB Certificated service and maintenance companies.

KEY TO ABBREVIATIONS

The following is a key to the abbreviations used in the column headed "Type" throughout the list.

Extinguishing Agent	Abbreviation
Water	W
Foam	F
Halogenated hydrocarbon	HH
Dry Powder	ABC or BC
Carbon Dioxide	CO_2

Method of Actuation

Stored pressure	(SP)
Gas cartridge	(GC)

Units
Under the column headed "Capacity", the following units and their abbreviations are used:
kg - kilogram
l   - litre

Operating Temperature Ranges
Each extinguisher listing includes details of the operating temperature range in accordance with EN 3-5: 1996. The operating temperature range options permitted by EN 3-5: 1996 are as follows:
Water and foam extinguishers: +5°C, 0°C, -10°C, -20°C, -25°C or -30°C to +60°C
Dry powder, carbon dioxide and halogenated hydrocarbon extinguishers: -20°C or -30°C to + 60°C.

Colour Coding
EN 3-5: 1996 permits a zone of colour of up to 5% of the external area of the extinguisher body to be used to identify the extinguishing agent. Each extinguisher listing details, by specification, the colour coding options available.

Recharging Details Category

Recharging to be undertaken by the manufacturer, his agent or an approved service and maintenance company. Section 2 details LPCB Certificated service and maintenance companies. | A

Reinstatement to be carried out by transferring the operating head to a replacement body. | B

Recharging to be carried out by the manufacturer only. | C

The body is not rechargeable and should be disposed of in accordance with the manufacturer's instructions. | D

Firemaster Extinguisher Limited

Firex House, 174-176 Hither Green Lane, London SE13 6QB

Tel: +44 (0)181 852 8585 • Fax: +44 (0)181 297 8020

Certificate No. 290a to EN 3-1 to 6
Approved products

Model	LPCB Ref. No.	Type	Capacity	Fire Rating	Operating Temp Range (°C)	Colour Coding	Recharging details	
							Category	Part
1000PR	290a/01	ABC(SP)	1kg	5A/34B	-20 to +60	BS 7863	A	P11119
PSP2000	290a/02	ABC(SP)	2kg	13A/70B	-20 to +60	BS 7863	A	P11120
FSP2000	290a/03	F(SP)	2l	8A/55B	+5 to +60	BS 7863	A	RF2
FSP2000/LT	290a/04	F(SP)	2l	8A/34B	-15 to +60	BS 7863	A	RF2/LT
PSP4000	290a/05	ABC(SP)	4kg	27A/113B	-20 to +60	BS 7863	A	P11121
PSP6000	290a/06	ABC(SP)	6kg	43A/183B	-20 to +60	BS 7863	A	P11122
PSP9000	290a/07	ABC(SP)	9kg	55A/233B	-20 to +60	BS 7863	A	P11123

Certificated products

Model	LPCB Ref. No.	Type	Capacity	Fire Rating	Operating Temp Range (°C)	Colour Coding	Recharging details	
WSP9000	290a/08	W(SP)	9l	13A	+5 to +60	BS 7863	A	-
FSP9000	290a/09	F(SP)	9l	27A/183B	+5 to +60	BS 7863	A	RF9

Walter Kidde Portable Equipment Inc

1394 South Third Street, Mebane, NC27302, USA

Tel: +1 919 563 5911 • Fax: +1 919 563 4582
UK Sales Enquiries: Tel: +44 (0)1844 218488

Certificate No. 406a to EN 3-1 to 6
Certificated products

Model	LPCB Ref. No.	Type	Capacity	Fire Rating	Operating Temp Range (°C)	Colour Coding	Recharging details	
							Category	Part
1 EN3 TC	406a/01	ABC(SP)	1kg	8A/34B	-20 to +60	BS 7863	D	-
2 EN3 TC	406a/02	ABC(SP)	2kg	13A/70B	-20 to +60	BS 7863	D	-

Angletex Fire Protection Limited
37/39 Manor Road, Romford, Essex RM1 2TL

Tel: +44 (0)1708 756248 • Fax: +44 (0)1708 769315

Product Conformity Cert. No. 229a
Assessed to BS 5306: Part 3: 1985

Certificated servicing and maintenance company authorised to use the LPCB Mark on, and issue Certificates of Conformity in respect of portable fire extinguishers serviced in accordance with BS 5306: Part 3: 1985.

Saffire Extinguishers Limited
Unit 16,Thames Trading Centre, Woodrow Way, off Fairhills Road, Irlam, Manchester M30 6BP

Tel: +44 (0)161 777 8346 • Fax: +44 (0)161 777 8347

Product Conformity Cert. No. 225a
Assessed to BS 5306: Part 3: 1985

Certificated servicing and maintenance company authorised to use the LPC Mark on, and issue Certificates of Conformity in respect of portable fire extinguishers serviced in accordance with BS 5306: Part 3: 1985.

Thameside Fire Protection
Unit 7, Swinbourne Court, Burnt Mills Industrial Estate, Basildon, Essex SS13 1QA.

Tel: +44 (0)1268 591059 • Fax: +44 (0)1268 590974

Product Conformity Cert. No. 269a
Assessed to BS 5306: Part 3: 1985

Certificated servicing and maintenance company authorised to use the LPCB Mark on, and issue Certificates of Conformity in respect of portable fire extinguishers serviced in accordance with BS 5306: Part 3: 1985.

Safe and Sound Fire Securities Limited
Unit 22, Thurlow Street, Stowell Technical Park, Salford, Manchester M5 2XH

Tel: +44 (0)161 736 0776 • Fax: +44 (0)161 736 0773

Product Conformity Cert. No. 392a
Assessed to BS 5306: Part 3: 1985

Certificated servicing and maintenance company authorised to use the LPC Mark on, and issue Certificates of Conformity in respect of portable fire extinguishers serviced in accordance with BS 5306: Part 3: 1985.

Part 5

AUTOMATIC SPRINKLER, WATER SPRAY AND DELUGE SYSTEMS

5 AUTOMATIC SPRINKLER, WATER SPRAY AND DELUGE SYSTEMS

INTRODUCTION

LPS 1048 Certificated Sprinkler Installers and Supervising Bodies (Section 1A)

Firms listed in Section 1A are LPS 1048 Certificated Sprinkler Installers authorised by the LPCB to design, install and certificate sprinkler installations to the *LPC Rules for Automatic Sprinkler Installations* (the 'LPC Rules'). These Rules incorporate BS 5306: Part 2: 1990 - *Fire extinguishing installations and equipment in premises: Specification for sprinkler systems.*

LPS 1048 Certificated Supervising Bodies are also identified in Section 1A, and are authorised by the LPCB to supervise LPS 1048 Registered Supervised Sprinkler Installers.

LPS 1048 Registered Supervised Sprinkler Installers (Section 1B)

Firms listed in Section 1B are LPS 1048 Registered Supervised Sprinkler Installers, recognised by the LPCB to design and install sprinkler installations to the LPC Rules under the supervision of an LPS 1048 Certificated Supervising Body.

LPS 1050 Certificated Sprinkler System Servicing and Maintenance Firms (Section 2A)

Firms listed in Section 2A are LPS 1050 Certificated Sprinkler System Servicing and Maintenance Firms. They are authorised to issue LPS 1050 Certificates of Servicing and Inspection for sprinkler systems serviced and maintained in accordance with BS 5306: Part 2: 1990.

LPS 1050 Certificated Supervising Bodies are also identified in Section 2A, and are authorised by the LPCB to supervise LPS 1050 Registered Supervised Sprinkler Servicing and Maintenance Firms.

LPCB Certificates of Conformity

LPS 1048 Certificated Sprinkler Installers shall issue LPS 1048 Certificates of Conformity for all completed installations, extensions and alterations conforming to the LPC Rules/ BS 5306: Part 2: 1990.

LPS 1048 Certificated Supervising Bodies shall issue LPS 1048 Certificates of Conformity in respect of all completed installations, extensions and alterations which are designed and installed by LPS 1048 Registered Supervised Sprinkler Installers.

LPS 1050 Certificated Sprinkler System Servicing firms shall issue LPS 1050 Certificates of Servicing and Inspection for all servicing and maintenance contracts conforming to BS 5306: part 2: 1990.

Additional Offices

Where a firm has additional offices, each office is required to meet the requirements of LPS 1048 and LPS 1050 to be eligible for certification, and is separately listed.

SPRINKLER EQUIPMENT (Section 3 onwards)

Equipment has been assessed and tested to the appropriate Standard, and the firms' manufacturing facilities have been certificated to ISO 9000. These facilities are re-inspected twice per annum and in addition the equipment is subject to re-examination and testing as necessary.

Where conformity with the LPC Rules is specified, LPCB certificated or approved equipment shall be used.

SPECIFIERS AND USERS PLEASE NOTE

1 LPCB Certificates of Conformity

Specifiers should demand LPCB Certificates of Conformity for all sprinkler contracts including extensions and alterations. This will ensure that contracts are carried out in accordance with the LPC Rules and LPS 1048 requirements.

2 Fully hydraulically calculated (FHC) sprinkler systems

The following firms are authorised under the LPS 1048 scheme to design and install fully hydraulically calculated sprinkler systems:

(a) All Certificated Installers (Section 1A).

(b) Only those Registered Supervised Installers (Section 1B) which have been approved by LPCB to carry out fully hydraulically calculated sprinkler systems. Check firm's status in list.

3 Financial assessment of LPS 1048 Installers

LPCB assess LPS 1048 Installers to the appropriate technical and quality standards.

LPCB **do not** carry out any financial assessment of LPS 1048 Installers with regard to their ability to finance contracts.

A & A Fire Limited

10-12 Glenfield Road, Kelvin Industrial Estate, East Kilbride, Lanarkshire G75 0RA

Tel: +44 (0)13552 37588 • Fax: +44 (0)13552 63399

Certificate No: CSI-030

Certificated Sprinkler Installer, authorised to use the LPCB Mark on, and issue LPCB Certificates of Conformity for, sprinkler systems conforming to the *LPC Rules for Automatic Sprinkler Installations,* incorporating BS 5306: Part 2: 1990, for all hazard classifications.

Authorised to use LPCB certificated and approved sprinkler heads and alarm valves manufactured by:
Central Sprinkler Corporation
Grinnell Manufacturing (UK)
Spraysafe Automatic Sprinklers Limited
The Reliable Automatic Sprinkler Co. Inc.

A & F Sprinklers Limited

574 Manchester Road, Blackford Bridge, Bury BL9 9SW

Tel: +44 (0)161 796 5397 • Fax: +44 (0)161 796 6057

Certificate No: CSI-004

Certificated Sprinkler Installer, authorised to use the LPCB Mark on, and issue LPCB Certificates of Conformity for, sprinkler systems conforming to the *LPC Rules for Automatic Sprinkler Installations,* incorporating BS 5306: Part 2: 1990, for all hazard classifications.

Authorized to use LPCB certificated and approved sprinkler heads and alarm valves manufactured by:
Angus Fire Armour Limited
Spraysafe Automatic Sprinklers Limited

Actspeed Limited

Unit 29, Silk Mill Estate, Brook Street, Tring, Hertfordshire HP23 5EF

Tel: +44 (0)1442 824442 • Fax: +44 (0)1442 827400

Certificate No: CSI-049

Certificated Sprinkler Installer, authorised to use the LPCB Mark on, and issue LPCB Certificates of Conformity for, sprinkler systems conforming to the *LPC Rules for Automatic Sprinkler Installations,* incorporating BS 5306: Part 2: 1990, for all hazard classifications.

Certificated Supervising Body, authorised to supervise Registered Sprinkler Installers.

Authorised to use LPCB certificated and approved sprinkler heads and alarm valves manufactured by:
Angus Fire Armour Limited.
The Reliable Automatic Sprinkler Co. Inc.
Viking Corporation.

Angus Fire Armour Limited

Thame Park Road, Thame, Oxfordshire OX9 3RT

Tel: +44 (0)1844 214545 • Fax: +44 (0)1844 213511

Certificate No: CSI-035

Certificated Sprinkler Installer, authorised to use the LPCB Mark on, and issue LPCB Certificates of Conformity for, sprinkler systems conforming to the *LPC Rules for Automatic Sprinkler Installations,* incorporating BS 5306: Part 2: 1990, for all hazard classifications.

Certificated Supervising Body, authorised to supervise Registered Supervised Sprinkler Installers.

Authorised to use LPCB certificated and approved sprinkler heads and alarm valves manufactured by:
Angus Fire Armour Limited

Argus Fire Protection Company Limited

Hendglade House, 46 New Road, Stourbridge, West Midlands DY8 1PA

Tel: +44 (0)1384 376256 • Fax: +44 (0)1384 393955

Certificate No: CSI-005

Certificated Sprinkler Installer, authorised to use the LPCB Mark on, and issue LPCB Certificates of Conformity for, sprinkler systems conforming to the *LPC Rules for Automatic Sprinkler Installations*, incorporating BS 5306: Part 2: 1990, for all hazard classifications.
Authorised to use LPCB certificated and approved sprinkler heads and alarm valves manufactured by:
    Grinnell Manufacturing (UK) Limited
    Spraysafe Automatic Sprinklers, USA

Armstrong Priestley Limited

Flaxton House, Greenmount Terrace, Beeston, Leeds LS11 6BX

Tel: +44 (0)113 2774499 • Fax: +44 (0)113 2760309

Certificate No: CSI-023

Certificated Sprinkler Installer, authorised to use the LPCB Mark on, and issue LPCB Certificates of Conformity for, sprinkler systems conforming to the *LPC Rules for Automatic Sprinkler Installations*, incorporating BS 5306 : Part 2 : 1990, for all hazard classifications.

Authorized to use LPCB Certificated and approved sprinkler heads and alarm valves manufactured by:
    Angus Fire Armour Limited
    Spraysafe Automatic Sprinklers Limited
    The Reliable Automatic Sprinkler Co. Inc.

Atlas Fire Engineering Limited

Unit House, 2-8 Morfa Road, Swansea SA1 2HS

Tel: +44 (0)1792 465006 • Fax: +44 (0)1792 648535

Certificate No: CSI-007

Certificated Sprinkler Installer, authorised to use the LPCB Mark on, and issue LPCB Certificates of Conformity for, sprinkler systems conforming to the *LPC Rules for Automatic Sprinkler Installations*, incorporating BS 5306 : Part 2 : 1990, for all hazard classifications.
Certificated Supervising Body, authorised to supervise Registered Supervised Sprinkler Installers.
Authorised to use LPCB certificated and approved sprinkler heads and alarm valves manufactured by:
    Grinnell Manufacturing (UK) Limited
    The Reliable Automatic Sprinkler Co Inc.
    Spraysafe Automatic Sprinklers Limited
    Viking Corporation

Besseges Limited

James Howe Mill, Turner Lane, Ashton-under-Lyne, Lancashire OL6 8LS

Tel: +44 (0)161 308 3252 • Fax: +44 (0)161 339 5003

Product Cert. No: CSI-043

Certificated Sprinkler Installer, authorised to use the LPCB Mark on, and issue Certificates of Conformity for, sprinkler systems conforming to the *LPC Rules for Automatic Sprinkler Installations*, incorporating BS 5306: Part 2: 1990, for all hazard classifications.
Authorised to use LPCB certificated and approved sprinkler heads and alarm valves manufactured by:
    Angus Fire Armour Limited
    Viking Corporation

5 SECTION 1A:
LPS 1048 Certificated Sprinkler Installers and Supervising Bodies

Central Fire Protection Limited
Unit 3, Peerglow Business Centre, Marsh Lane, Ware, Herts SG12 9QL

Tel: +44 (0)1920 460616 • Fax: +44 (0)1920 461113

Certificate No: CSI-026

Certificated Sprinkler Installer, authorised to use the LPCB Mark on, and issue LPCB Certificates of Conformity for, sprinkler systems conforming to the *LPC Rules for Automatic Sprinkler Installations,* incorporating BS 5306: Part 2: 1990, for all hazard classifications.
Authorised to use LPCB certificated and approved sprinkler heads and alarm valves manufactured by:
    Grinnell Manufacturing (UK) Limited
    The Reliable Automatic Sprinkler Co. Inc.
    Viking Corporation

Dalkia Technical Services Limited
7 Twisleton Court, Priory Hill, Dartford, Kent DA1 2EN

Tel: +44 (0)1322 226600 • Fax: +44 (0)1322 287819

Certificate No: CSI-014

Certificated Sprinkler Installer, authorised to use the LPCB Mark on, and issue LPCB Certificates of Conformity for, sprinkler systems conforming to the *LPC Rules for Automatic Sprinkler Installations,* incorporating BS 5306: Part 2: 1990, for all hazard classifications.
Authorised to use LPCB certificated and approved sprinkler heads and alarm valves manufactured by:
    The Reliable Automatic Sprinkler Co. Inc.

Fire Control (Glasgow) Limited
2 Albion Way, Kelvin Industrial Estate, East Kilbride, G75 0YN

Tel: +44 (0)13552 37125 • Fax: +44 (0)13552 64359

Product Cert. No: CSI-046

Certificated Sprinkler Installer, authorised to use the LPCB Mark on, and issue LPCB Certificates of Conformity for, sprinkler systems conforming to the LPC Rules for Automatic Sprinkler Installations, incorporating BS 5306: Part 2: 1990, for all hazard classifications.
Authorised to use LPCB certificated and approved sprinkler heads and alarm valves manufactured by:
    Grinnell Manufacturing (UK) Limited
    The Reliable Automatic Sprinkler Co. Inc.
    Spraysafe Automatic Sprinklers Limited
    Viking Corporation

Fire Defender (UK) Limited
Meridian Centre, King Street, Oldham, Lancashire OL8 1EZ

Tel: +44 (0)161 678 0125 • Fax: +44 (0)161 627 4825

Certificate No: CSI-002

Certificated Sprinkler Installer, authorised to use the LPCB Mark on, and issue LPCB Certificates of Conformity for, sprinkler systems conforming to the *LPC Rules for Automatic Sprinkler Installations,* incorporating BS 5306: Part 2: 1990, for all hazard classifications.
Authorised to use LPCB certificated and approved sprinkler heads and alarm valves manufactured by:
    Grinnell Manufacturing (UK) Limited

Fireproof Fire Engineering Limited

The Old Brewery, Coldhurst Street, Oldham, Lancashire OL1 2BQ

Tel: +44 (0)161 620 0902 • Fax: +44 (0)161 620 0772

Certificate No: CSI-029

Certificated Sprinkler Installer, authorised to use the LPCB Mark on, and issue LPCB Certificates of Conformity for, sprinkler systems conforming to the *LPC Rules for Automatic Sprinkler Installations*, incorporating BS 5306: Part 2: 1990, for all hazard classifications.
Authorised to use LPCB certificated and approved sprinkler heads and alarm valves manufactured by:
Angus Fire Armour Limited
Grinnell Manufacturing (UK) Limited
Spraysafe Automatic Sprinklers Limited
Viking Corporation

Fire Security (Sprinkler Installations) Limited

Homefield Road, Haverhill, Suffolk CB9 8QP

Tel: +44 (0)1440 705815 • Fax: +44 (0)1440 704352

Certificate No: CSI-025

Certificated Sprinkler Installer, authorised to use the LPCB Mark on, and issue LPCB Certificates of Conformity for, sprinkler systems conforming to the *LPC Rules for Automatic Sprinkler Installations*, incorporating BS 5306: Part 2: 1990, for all hazard classifications.
Authorised to use LPCB certificated and approved sprinkler heads and alarm valves manufactured by:
Spraysafe Automatic Sprinklers Limited
Viking Corporation

Grinnell Firekil

Firekil House, 37 Mark Road, Hemel Hempstead, Hertfordshire HP2 7BW

Tel: +44 (0)1442 232344 • Fax: +44 (0)1442 217623

Certificate No: CSI-016

Certificated Sprinkler Installer, authorised to use the LPCB Mark on, and issue LPCB Certificates of Conformity for, sprinkler systems conforming to the *LPC Rules for Automatic Sprinkler Installations*, incorporating BS 5306: Part 2: 1990, for all hazard classifications.
Authorised to use LPCB certificated and approved sprinkler heads and alarm valves manufactured by:
Grinnell Corporation
Grinnell Manufacturing (UK) Limited
The Reliable Automatic Sprinkler Co. Inc.

Haden Fire Protection (A Division of Haden Young Limited)

Britannia Road, Patchway, Bristol BS12 5TD

Tel: +44 (0)117 9693911 • Fax: +44 (0)117 9798711

Product Cert. No: CSI-044

Certificated Sprinkler Installer, authorised to use the LPCB Mark on, and issue LPCB Certificates of Conformity for, sprinkler systems conforming to the *LPC Rules for Automatic Sprinkler Installations*, incorporating BS 5306: Part 2: 1990, for all hazard classifications.
Authorised to use LPCB certificated and approved sprinkler heads and alarm valves manufactured by:
Angus Fire Armour Limited
Grinnell Manufacturing (UK) Limited
The Reliable Automatic Sprinkler Co. Inc.
Viking Corporation.

5 SECTION 1A:

LPS 1048 CERTIFICATED SPRINKLER INSTALLERS AND SUPERVISING BODIES

Haden Young Fire Engineering

Iveco Ford House, Station Road, Watford WD1 1HB

Tel: +44 (0)1923 254646 • Fax: +44 (0)1923 296300

Certificate No: CSI-040

Certificated Sprinkler Installer, authorised to use the LPCB Mark on, and issue LPCB Certificates of Conformity for, sprinkler systems conforming to the *LPC Rules for Automatic Sprinkler Installations,* incorporating BS 5306: Part 2: 1990, for all hazard classifications.
Certificated Supervising Body, authorised to supervise Registered Supervised Sprinkler Installers
Authorised to use LPCB certificated and approved sprinkler heads and alarm valves manufactured by:
    Spraysafe Automatic Sprinklers Limited
    Viking Corporation

Hall and Kay Fire Engineering

Sterling Park, Clapgate Lane, Woodgate Valley, West Midlands B32 3BU

Tel: +44 (0)121 421 3311 • Fax: +44 (0)121 422 7312

Certificate No: CSI-003

Certificated Sprinkler Installer, authorised to use the LPCB Mark on, and issue LPCB Certificates of Conformity for, sprinkler systems conforming to the *LPC Rules for Automatic Sprinkler Installations,* incorporating BS 5306: Part 2: 1990, for all hazard classifications.
Certificated Supervising Body, authorised to supervise Registered Supervised Sprinkler Installers.
Authorised to use LPCB certificated and approved sprinkler heads and alarm valves manufactured by:
    G. W. Sprinkler A/S
    Spraysafe Automatic Sprinklers Limited

Hall and Kay Fire Engineering

10 Westpoint, Clarence Avenue, Trafford Park, Manchester M17 1QS

Tel: +44 (0)161 872 7316 • Fax: +44 (0)161 872 7003

Certificate No: CSI-032

Certificated Sprinkler Installer, authorised to use the LPCB Mark on, and issue LPCB Certificates of Conformity for, sprinkler systems conforming to the *LPC Rules for Automatic Sprinkler Installations,* incorporating BS 5306: Part 2: 1990, for all hazard classifications.
Authorised to use LPCB certificated and approved sprinkler heads and alarm valves manufactured by:
    G. W. Sprinkler A/S
    Spraysafe Automatic Sprinklers Limited

Hall and Kay Fire Engineering

64 Mill Street, Slough, Berkshire SL2 5DH

Tel: +44 (0)1753 554500 • Fax: +44 (0)1753 554616

Certificate No CSI-052

Certificated Sprinkler Installer, authorised to use the LPCB Mark on, and issue LPCB Certificates of Conformity for, sprinkler systems conforming to the LPC Rules for Automatic Sprinkler Installations, incorporating BS 5306: Part 2: 1990, for all hazard classifications.
Authorised to use LPCB Certificated and approved sprinkler heads and alarm valves manufactured by:
    Spraysafe Automatic Sprinklers Limited
    G W Sprinkler A/S

Hall Fire Protection Limited
186 Moorside Road, Swinton, Manchester M27 9HA.
Also at: Delta Works, Chadwick Road, Eccles, Lancashire M30 0NZ

Tel: +44 (0)161 793 4822 • Fax: +44 (0)161 794 4950

Certificate No: CSI-012

Certificated Sprinkler Installer, authorised to use the LPCB Mark on, and issue Certificates of Conformity for, sprinkler systems conforming to the *LPC Rules for Automatic Sprinkler Installations,* incorporating BS 5306: Part 2: 1990, for all hazard classifications.
Certificated Supervising Body, authorised to supervise Registered Supervised Sprinkler Installers.
Authorised to use LPCB certificated and approved sprinkler heads and alarm valves manufactured by:
    Figgie Fire Protection Systems
    Grinnell Manufacturing (UK) Limited
    GW Sprinkler A/S
    Spraysafe Automatic Sprinklers Limited
    Viking Corporation

How Fire Limited
Hillcrest Business Park, Cinderbank, Dudley, West Midlands DY2 9AP

Tel: +44 (0)1384 458993 • Fax: +44 (0)1384 458981

Certificate No: CSI-001

Certificated Sprinkler Installer, authorised to use the LPCB Mark on, and issue LPCB Certificates of Conformity for, sprinkler systems conforming to the *LPC Rules for Automatic Sprinkler Installations,* incorporating BS 5306: Part 2: 1990, for all hazard classifications.
Authorised to use LPCB certificated and approved sprinkler heads and alarm valves manufactured by:
    Grinnell Manufacturing (UK) Limited
    The Reliable Automatic Sprinkler Co. Inc.
    Spraysafe Automatic Sprinklers Limited
    Viking Corporation

Irish Sprinkler & Fire Protection Limited
Waterways House, Grand Canal Quay, Dublin 2, Ireland

Tel: +44 (0)1 671 1500 • Fax: +44 (0)1 671 1600

Certificate No: CSI-051

Certificated Sprinkler Installer, authorised to use the LPCB Mark on, and issue LPCB Certificates of Conformity for, sprinkler systems conforming to the *LPC Rules for Automatic Sprinkler Installations,* incorporating BS 5306: Part 2: 1990, for all hazard classifications.
Authorised to use LPCB certificated and approved sprinkler heads and alarm valves manufactured by:
    Angus Fire Armour Limited
    The Reliable Automatic Sprinkler Co. Inc.

L & S Building Services Limited
Fir Tree Place, 77 Church Road, Ashford, Middlesex TW15 2PE

Tel: +44 (0)1784 245062 • Fax: +44 (0)1784 247368

Certificate No: CSI-031

Certificated Sprinkler Installer, authorised to use the LPCB Mark on, and issue LPCB Certificates of Conformity for, sprinkler systems conforming to the *LPC Rules for Automatic Sprinkler Installations,* incorporating BS 5306: Part 2: 1990, for all hazard classifications.
Authorised to use LPCB certificated and approved sprinkler heads and alarm valves manufactured by:
    Angus Fire Armour Limited

Linkester Fire Protection Limited
4/6 Cross Street, Macclesfield, Cheshire SK11 7PG

Tel: +44 (0)1625 511272 • Fax: +44 (0)1625 511272

Certificate No: CSI-021

Certificated Sprinkler Installer, authorised to use the LPCB Mark on, and issue LPCB Certificates of Conformity for, sprinkler systems conforming to the *LPC Rules for Automatic Sprinkler Installations,* incorporating BS 5306: Part 2: 1990, for all hazard classifications.
Authorised to use LPCB certificated and approved sprinkler heads and alarm valves manufactured by:
Spraysafe Automatic Sprinklers Limited
Viking Corporation

Mather & Platt (Ireland) Limited
7 Ardee Road, Rathmines, Dublin 6, Ireland

Tel: +353 4966077 • Fax: +353 4966858

Certificate No: CSI-047

Certificated Sprinkler Installer, authorised to use the LPCB Mark on, and issue LPCB Certificates of Conformity for, sprinkler systems conforming to the *LPC Rules for Automatic Sprinkler Installations,* incorporating BS 5306: Part 2: 1990, for all hazard classifications.
Certificated Supervising Body, authorised to supervise Registered Supervised Sprinkler Installers.
Authorised to use LPCB certificated and approved sprinkler heads and alarm valves manufactured by:
Grinnell Corporation
Grinnell Manufacturing (UK) Limited

Matthew Hall Limited
Fire and Communication Systems
7-14 Great Dover Street, London SE1 4YR

Tel: +44 (0)171 407 7272 • Fax: +44 (0)171 403 6390

Certificate No: CSI-015

Certificated Sprinkler Installer, authorised to use the LPCB Mark on, and issue LPCB Certificates of Conformity for, sprinkler systems conforming to the *LPC Rules for Automatic Sprinkler Installations,* incorporating BS 5306: Part 2: 1990, for all hazard classifications.
Authorised to use LPCB certificated and approved sprinkler heads and alarm valves manufactured by:
Angus Fire Armour Limited
Grinnell Manufacturing (UK) Limited
The Reliable Automatic Sprinkler Co. Inc.
Spraysafe Automatic Sprinklers Limited
Viking Corporation

Matthew Hall Limited
Fire and Communications Systems
Unit 1A, Stag Industrial Estate, Oxford Street, Bilston, West Midlands WV14 7HR

Tel: +44 (0)1902 493100 • Fax: +44 (0)1902 405922

Product Cert. No: CSI-045

Certificated Sprinkler Installer, authorised to use the LPCB Mark on, and issue LPCB Certificates of Conformity for, sprinkler systems conforming to the *LPC Rules for Automatic Sprinkler Installations,* incorporating BS 5306: Part 2: 1990, for all hazard classifications.
Authorised to use LPCB certificated and approved sprinkler heads and alarm valves manufactured by:
Angus Fire Armour Limited
Grinnell Manufacturing (UK) Limited
The Reliable Automatic Sprinkler Co. Inc.
Spraysafe Automatic Sprinklers Limited
Viking Corporation

Mercury Engineering Limited

Mercury House, Sandyford Industrial Estate, Foxrock, Dublin 18, Ireland

Tel: +353 295 2211 • Fax: +353 295 3202

Certificate No: CSI-048

Certificated Sprinkler Installer, authorised to use the LPCB Mark on, and issue LPCB Certificates of Conformity for, sprinkler systems conforming to the *LPC Rules for Automatic Sprinkler Installations*, incorporating BS 5306: Part 2: 1990, for all hazard classifications.
Certificated Supervising Body, authorised to supervise Registered Supervised Sprinkler Installers.
Authorised to use LPCB certificated and approved sprinkler heads and alarm valves manufactured by:
    Angus Fire Armour Limited
    Grinnell Corporation
    The Reliable Automatic Sprinkler Co. Inc.
    Spraysafe Automatic Sprinklers Limited

Olympic Fire Protection Limited

10 Stratfield Park, Waterlooville, Hampshire PO7 7XN

Tel: +44 (0)1705 231444 • Fax: +44 (0)1705 250666

Certificate No: CSI-050

Certificated Sprinkler Installer, authorised to use the LPCB Mark on, and issue LPCB Certificates of Conformity for, sprinkler systems conforming to the *LPC Rules for Automatic Sprinkler Installations*, incorporating BS 5306: Part 2: 1990, for all hazard classifications.
Authorised to use LPCB certificated and approved sprinkler heads and alarm valves manufactured by:
    Angus Fire Armour Limited
    The Reliable Automatic Sprinkler Co. Inc.
    Viking Corporation

Pendle Designs Limited

16-20 Heaton Fold, Fishpool, Bury, Lancashire BL9 9HF

Tel: +44 (0)161 764 2252 • Fax: +44 (0)161 763 1429

Certificate No: CSI-027

Certificated Sprinkler Installer, authorised to use the LPCB Mark on, and issue LPCB Certificates of Conformity for, sprinkler systems conforming to the *LPC Rules for Automatic Sprinkler Installations*, incorporating BS 5306: Part 2: 1990, for all hazard classifications.
Authorised to use LPCB certificated and approved sprinkler heads and alarm valves manufactured by:
    Angus Fire Armour Limited
    Grinnell Manufacturing (UK) Limited
    Spraysafe Automatic Sprinklers Limited

Preussag Fire Protection Limited

Field Way, Greenford, Middlesex UB6 8UZ

Tel: +44 (0)181 832 2000 • Fax: +44 (0)181 832 2200

Certificate No: CSI-034

Certificated Sprinkler Installer, authorised to use the LPCB Mark on, and issue LPCB Certificates of Conformity for, sprinkler systems conforming to the *LPC Rules for Automatic Sprinkler Installations*, incorporating BS 5306: Part 2: 1990, for all hazard classifications.
Certificated Supervising Body, authorised to supervise Registered Supervised Sprinkler Installers.
Authorised to use LPCB certificated and approved sprinkler heads and alarm valves manufactured by:
    Minimax GmbH.
    Viking Corporation

SECTION 1A:

LPS 1048 CERTIFICATED SPRINKLER INSTALLERS AND SUPERVISING BODIES

Preussag Fire Protection Limited

Mersey House, 220 Stockport Road, Cheadle Heath, Stockport SK3 0LX

Tel: +44 (0)161 428 3661 • Fax: +44 (0)161 428 3662

Certificate No CSI-053

Certificated Sprinkler Installer, authorised to use the LPCB Mark on, and issue LPCB Certificates of Conformity for, sprinkler systems conforming to the LPC Rules for Automatic Sprinkler Installations, incorporating BS 5306: Part 2: 1990, for all hazard classifications.
Authorised to use LPCB Certificated and approved sprinkler heads and alarm valves manufactured by:
  Minimax GmbH
  Viking Corporation

Project Fire Engineers Limited

Sandyford Street, Stafford ST16 3NF

Tel: +44 (0)1785 222999 • Fax: +44 (0)1785 222959

Certificate No: CSI-011

Certificated Sprinkler Installer, authorised to use the LPCB Mark on, and issue LPCB Certificates of Conformity for, sprinkler systems conforming to the *LPC Rules for Automatic Sprinkler Installations,* incorporating BS 5306: Part 2: 1990, for all hazard classifications.
Certificated Supervising Body, authorised to supervise Registered Supervised Sprinkler Installers.
Authorised to use LPCB certificated and approved sprinkler heads and alarm valves manufactured by:
  Central Sprinkler Corporation

Ross Fire Protection Limited

29 Deerdykes View, Westfield, Cumbernauld G68 9HN

Tel: +44 (0)1236 738502 • Fax: +44 (0)1236 727977

Certificate No: CSI-036

Certificated Sprinkler Installer, authorised to use the LPCB Mark on, and issue LPCB Certificates of Conformity for, sprinkler systems conforming to the *LPC Rules for Automatic Sprinkler Installations,* incorporating BS 5306: Part 2: 1990, for all hazard classifications.
Authorised to use LPCB certificated and approved sprinkler heads and alarm valves manufactured by:
  Angus Fire Armour Limited

Rotary Firematic Limited

5 Trench Road, Mallusk Industrial Estate, Newtownabbey, Co Antrim, Northern Ireland BT36 8XA

Tel: +44 (0)1232 831200 • Fax: +44 (0)1232 831201

Certificate No: CSI-028

Certificated Sprinkler Installer, authorised to use the LPCB Mark on, and issue LPCB Certificates of Conformity for, sprinkler systems conforming to the *LPC Rules for Automatic Sprinkler Installations,* incorporating BS 5306: Part 2: 1990, for all hazard classifications.
Authorised to use LPCB certificated and approved sprinkler heads and alarm valves manufactured by:
  Angus Fire Armour Limited
  Grinnell Manufacturing (UK) Limited

TPT Fire Protection Services Limited
Plantagenet House, Kingsclere Park, Kingsclere, Newbury RG15 8SW

Tel: +44 (0)1635 299014 • Fax: +44 (0)1635 299727

Certificate No: CSI-041

Certificated Sprinkler Installer, authorised to use the LPCB Mark on, and issue LPCB Certificates of Conformity for, sprinkler systems conforming to the *LPC Rules for Automatic Sprinker Installations*, incorporating BS 5306: Part 2: 1990, for all hazard classifications.
Authorised to use LPCB certificated and approved sprinkler heads and alarm valves manufactured by:
    Angus Fire Armour Limited
    The Reliable Automatic Sprinkler Co. Inc.

Wormald Ansul (UK) Limited, Wormald Fire Systems,
Wormald Park, Grimshaw Lane, Newton Heath, Manchester M40 2WL

Tel: +44 (0)161 205 2321 • Fax: +44 (0)161 455 4459

Certificate No: CSI-008

Certificated Sprinkler Installer, authorised to use the LPCB Mark on, and issue LPCB Certificates of Conformity for, sprinkler systems conforming to the *LPC Rules for Automatic Sprinker Installations*, incorporating BS 5306: Part 2: 1990, for all hazard classifications.
Certificated Supervising Body, authorised to supervise Registered Supervised Sprinkler Installers.
Authorised to use LPCB certificated and approved sprinkler heads and alarm valves manufactured by:
    Grinnell Corporation
    Grinnell Manufacturing (UK) Limited
    The Reliable Automatic Sprinkler Co. Inc.
    Spraysafe Automatic Sprinklers Limited

Wormald Ansul (UK) Limited, Wormald Fire Systems
Hawthorns Business Centre, Halfords Lane, Smethwick, Warley,
West Midlands B66 1DW

Tel: +44 (0)121 558 4817 • Fax: +44 (0)121 558 2282

Certificate No: CSI-009

Certificated Sprinkler Installer, authorised to use the LPCB Mark on, and issue LPCB Certificates of Conformity for, sprinkler systems conforming to the *LPC Rules for Automatic Sprinkler Installations*, incorporating BS 5306: Part 2: 1990, for all hazard classifications.
Certificated Supervising Body, authorised to supervise Registered Supervised Sprinkler Installers.
Authorised to use LPCB certificated and approved sprinkler heads and alarm valves manufactured by:
    Grinnell Corporation
    Grinnell Manufacturing (UK) Limited
    The Reliable Automatic Sprinkler Co. Inc.
    Spraysafe Automatic Sprinklers Limited

Wormald Ansul (UK) Limited
Lowe Industrial Estate, 31 Ballynahinch Road, Carryduff, Belfast, Northern Ireland BT8 8EH

Tel: +44 (0)1232 813699 • Fax: +44 (0)1232 814418

Certificate No: CSI-042

Certificated Sprinkler Installer, authorised to use the LPCB Mark on, and issue LPCB Certificates of Conformity for, sprinkler systems conforming to the *LPC Rules for Automatic Sprinkler Installations*, incorporating BS 5306: Part 2: 1990, for all hazard classifications.
Authorised to use LPCB certificated and approved sprinkler heads and alarm valves manufactured by:
    Grinnell Corporation
    Grinnell Manufacturing (UK) Limited

Wormald Ansul (UK) Limited, Wormald Fire Systems
1A Unity Street, Bristol BS2 0HX

Tel: +44 (0)117 9277271 • Fax: +44 (0)117 9297586

Certificate No: CSI-010

Certificated Sprinkler Installer, authorised to use the LPCB Mark on, and issue LPCB Certificates of Conformity for, sprinkler systems conforming to the *LPC Rules for Automatic Sprinkler Installations,* incorporating BS 5306: Part 2: 1990, for all hazard classifications.
Authorised to use LPCB certificated and approved sprinkler heads and alarm valves manufactured by:
  Grinnell Corporation
  Grinnell Manufacturing (UK) Limited
  The Reliable Automatic Sprinkler Co. Inc.
  Spraysafe Automatic Sprinklers Limited

Wormald Ansul (UK) Limited, Wormald Fire Systems
30-32 Singer Road, Kelvin Industrial Estate, East Kilbride, G75 0XS

Tel: +44 (0)13552 25132 • Fax: +44 (0)13552 49268

Certificate No: CSI-018

Certificated Sprinkler Installer, authorised to use the LPCB Mark on, and issue LPCB Certificates of Conformity for, sprinkler systems conforming to the *LPC Rules for Automatic Sprinkler Installations,* incorporating BS 5306: Part 2: 1990, for all hazard classifications.
Authorised to use LPCB certificated and approved sprinkler heads and alarm valves manufactured by:
  Grinnell Corporation
  Grinnell Manufacturing (UK) Limited
  The Reliable Automatic Sprinkler Co. Inc.
  Spraysafe Automatic Sprinklers Limited

Wormald Ansul (UK) Limited, Wormald Fire Systems
205-206 Bedford Avenue, Slough Trading Estate, Slough, Berkshire SL1 4RY

Tel: +44 (0)1753 574111 • Fax: +44 (0)1753 824226

Certificate No: CSI-019

Certificated Sprinkler Installer, authorised to use the LPCB Mark on, and issue LPCB Certificates of Conformity for, sprinkler systems conforming to the *LPC Rules for Automatic Sprinkler Installations,* incorporating BS 5306: Part 2: 1990, for all hazard classifications.
Certificated Supervising Body, authorised to supervise Registered Supervised Sprinkler Installers.
Authorised to use LPCB certificated and approved sprinkler heads and alarm valves manufactured by:
  Grinnell Corporation
  Grinnell Manufacturing (UK) Limited
  The Reliable Automatic Sprinkler Co. Inc.
  Spraysafe Automatic Sprinklers Limited

Active Fire Protection Limited ✔

1st Floor, Hodsons House, Stafford Road, Penkridge, Staffs ST19 5AS

Tel: +44 (0)1785 716100 • Fax: +44 (0)1785 715151

Certificate No : RSI-038

Registered Supervised Sprinkler Installer, recognised to design and install, under the supervision of an LPCB Certificated Supervising Body, sprinkler systems conforming to the *LPC Rules for Automatic Sprinkler Installations*, incorporating BS 5306: Part 2: 1990.
Supervising Company: Angus Fire Armour Limited

A J Sprinkler & Pipefitting Company Limited

51 Queens Crescent, Gorleston, Great Yarmouth, Norfolk NR31 7NN*
*Also at 227 Hazeldene Road, Northampton NN2 7NZ

Tel: +44 (0)1493 663190 • Fax: +44 (0)1493 442030

Certificate No. RSI-076

Registered Supervised Sprinkler Installer, recognised to design and install, under the supervision of an LPCB Certificated Supervising Body, sprinkler systems conforming to the LPC Rules for Automatic Sprinkler Installations incorporating BS 5306 : Part 2 : 1990.
Supervising Company : Angus Fire Armour Limited

Alpine Fire Engineers Limited

236 Halifax Road, Smallbridge, Rochdale, Lancashire OL16 2NT

Tel: +44 (0)1706 357040 • Fax: +44 (0)1706 355924

Certificate No : RSI-049

Registered Supervised Sprinkler Installer, recognised to design and install, under the supervision of an LPCB Certificated Supervising Body, sprinkler systems conforming to the *LPC Rules for Automatic Sprinkler Installations*, incorporating BS 5306: Part 2: 1990.
Supervising Company: Angus Fire Armour Limited

Angletex Fire Protection Limited ✔

37/39 Manor Road, Romford, Essex RM1 2TL

Tel: +44 (0)1708 756248 • Fax: +44 (0)1708 769315

Certificate No: RSI-009

Registered Supervised Sprinkler Installer, recognised to design and install, under the supervision of an LPCB Certificated Supervising Body, sprinkler systems conforming to the *LPC Rules for Automatic Sprinkler Installations*, incorporating BS5306 : Part 2 : 1990.
Supervising Company: Angus Fire Armour Limited

Ashvale Fire Services Limited

Ashvale House, Fitzroy Street, Ashton-Under-Lyne, Lancashire OL7 0JG

Tel: +44 (0)161 343 1090 • Fax: +44 (0)161 343 3732

Certificate No: RSI-029

Registered Supervised Sprinkler Installer, recognised to design and install, under the supervision of an LPCB Certificated Supervising Body, sprinkler systems conforming to the *LPC Rules for Automatic Sprinkler Installations*, incorporating BS 5306: Part 2: 1990.
Supervising Company: Angus Fire Armour Limited

Automatic Fire Control Limited

Unit 9, Dorcan Business Village, Murdoch Road, Dorcan, Swindon SN3 5HY

Tel: +44 (0)1793 496624 • Fax: +44 (0)1793 496623

Certificate No: RSI-012

Registered Supervised Sprinkler Installer, recognised to design and install, under the supervision of an LPCB Certificated Supervising Body, sprinkler systems conforming to the *LPC Rules for Automatic Sprinkler Installations,* incorporating BS 5306 : Part 2 : 1990.
Supervising Company: Actspeed Limited
Hall Fire Protection Limited

B & C Building Services Limited

Sidney House, Aylestone Lane, Wigston, Leicester LE18 1BD.

Tel: +44 (0)116 288 2993 • Fax: +44 (0)116 288 2380

Certificate No : RSI-044

Registered Supervised Sprinkler Installer, recognised to design and install, under the supervision of an LPCB Certificated Supervising Body, sprinkler systems conforming to the LPC Rules for Automatic Sprinkler Installations, incorporating BS 5306: Part 2: 1990.
Supervising Company: Angus Fire Armour Limited

Bush Engineering

134 Biggin Hill, Upper Norwood, London SE19 3HP

Tel: +44 (0)181 679 9997 • Fax: +44 (0)181 764 6179

Certificate No: RSI-068

Registered Supervised Sprinkler Installer, recognised to design and install, under the supervision of an LPCB Certificated Supervising Body, sprinkler systems conforming to the *LPC Rules for Automatic Sprinkler Installations,* incorporating BS 5306: Part 2: 1990.
Supervising Company: Angus Fire Armour Limited

Cador Fire Protection Services Limited

Cuba Industrial Estate, Ramsbottom, Bury, Lancashire, BL0 0NE

Tel: +44 (0)1706 825111 • Fax: +44 (0)1706 826333

Certificate No: RSI-026

Registered Supervised Sprinkler Installer, recognised to design and install, under the supervision of an LPCB Certificated Supervising Body, sprinkler systems conforming to the *LPC Rules for Automatic Sprinker Installations,* incorporating BS 5306: Part 2: 1990.
Supervising Company: Actspeed Limited.

Compco Fire Systems Limited ✔

Holmer House, 26 London Road, Worcester WR5 2DL

Tel: +44 (0)1905 763800 • Fax: +44 (0)1905 763820

Certificate No: RSI-043

Registered Supervised Sprinkler Installer, recognised to design and install, under the supervision of an LPCB Certificated Supervising Body, sprinkler systems conforming to the *LPC Rules for Automatic Sprinker Installations,* incorporating BS 5306: Part 2: 1990.
Supervising Company: Angus Fire Armour Limited.

CPM Contract Services Limited
170-180 Carlton Road, Nottingham NG3 2BB

Tel: +44 (0)115 911 8844 • Fax: +44 (0)115 941 0923

Certificate No: RSI-067
Registered Supervised Sprinkler Installer, recognised to design and install, under the supervision of an LPCB
Certificated Supervising Body, sprinkler systems conforming to the LPC Rules for Automatic Sprinkler Installations,
incorporating BS 5306 : Part 2 : 1990.
Supervising Companies: Angus Fire Armour Limited

Crown House Engineering
Stephenson House, 2 Cherry Orchard Road, Croydon CR0 6BA

Tel: +44 (0)181 662 4700 • Fax: +44 (0)181 662 4795

Certificate No: RSI-014
Registered Supervised Sprinkler Installer, recognised to design and install, under the supervision of an LPCB
Certificated Supervising Body, sprinkler systems conforming to the *LPC Rules for Automatic Sprinkler Installations,*
incorporating BS 5306: Part 2: 1990.
Supervising Company: Angus Fire Armour Limited

Cutbill Fire Protection Services Limited
Trading as: Cutbill Fire Systems
Derwent House, Mary Anne Street, St. Paul's Square, Birmingham B3 1RL

Tel: +44 (0)121 212 9822 • Fax: +44 (0)121 233 2206

Certificate No: RSI-025
Registered Supervised Sprinkler Installer, recognised to design and install, under the supervision of an LPCB
Certificated Supervising Body, sprinkler systems conforming to the *LPC Rules for Automatic Sprinker Installations,*
incorporating BS 5306: Part 2: 1990.
Supervising Company: Angus Fire Armour Limited
Hall Fire Protection Limited.

Davencrest Limited
Manlec Building, Tudor Trading Estate, Ashton Street, Dukinfield,
Cheshire SK16 4RR

Tel: +44 (0)161 339 1550 • Fax: +44 (0)161 339 1550

Certificate No: RSI-055
Registered Supervised Sprinkler Installer, recognised to design and install, under the supervision of an LPCB
Certificated Supervising Body, sprinkler systems conforming to the LPC Rules for Automatic Sprinkler Installations,
incorporating BS 5306 : Part 2 : 1990.
Supervising Company: Angus Fire Armour Limited.

D D H Sprinklers Limited
31 Cherry Orchard Industrial Estate, Dublin 10, Ireland

Tel: +353 1 626 7970 • Fax: + 353 1 626 5873

Certificate no: RSI-039
Registered Supervised Sprinkler Installer, recognised to design and install, under the supervision of an LPCB
Certificated Supervising Body, sprinkler systems conforming to the LPC Rules for Automatic Sprinkler Installations,
incorporating BS 5306 : Part 2 : 1990.
Supervising Company: Angus Fire Armour Limited.

5 SECTION 1B:

LPS 1048 Registered Supervised Sprinkler Installers

D.I.S. Sprinklers

183 Westgate Street, Gloucester GL1 2RW

Tel: +44 (0)1452 330585 • Fax: +44 (0)1452 300195

Certificate No: RSI-010

Registered Supervised Sprinkler Installer, recognised to design and install, under the supervision of an LPCB Certificated Supervising Body, sprinkler systems conforming to the *LPC Rules for Automatic Sprinker Installations*, incorporating BS 5306: Part 2: 1990.
Supervising Company: Angus Fire Armour Limited

Elliott Fire Protection Limited

Unit 11, Yew Tree Industrial Estate, Mill Hall, Aylesford, Kent ME20 7ET

Tel: +44 (0)1622 716126 • Fax: +44 (0)1622 715007

Certificate No: RSI-063

Registered Supervised Sprinkler Installer, recognised to design and install, under the supervision of an LPCB Certificated Supervising Body, sprinkler systems conforming to the *LPC Rules for Automatic Sprinker Installations*, incorporating BS 5306: Part 2: 1990.
Supervising Company: Angus Fire Armour Limited

Fire Defence plc

Unit 6, Station Road, South Molton, Devon EX36 3LL

✔

Tel: +44 (0)1769 574070 • Fax: +44 (0)1769 574079

Certificate No: RSI-037

Registered Supervised Sprinkler Installer, recognised to design and install, under the supervision of an LPCB Certificated Supervising Body, sprinkler systems conforming to the *LPC Rules for Automatic Sprinkler Installations*, incorporating BS 5306: Part 2: 1990.

Supervising Company: Angus Fire Armour Limited.

Fire Protection & Alarm Limited

23 Langlands Place, Kelvin South Business Park, East Kilbride Lanarkshire G75 0YF

Tel: +44 (0)13552 60222 • Fax: +44 (0)13552 60444

Certificate No: RSI-016

Registered Supervised Sprinkler Installer, recognised to design and install, under the supervision of an LPCB Certificated Supervising Body, sprinkler systems conforming to the *LPC Rules for Automatic Sprinkler Installations*, incorporating BS 5306: Part 2: 1990.
Supervising Company: Hall Fire Protection Limited

Fire Sprinkler Services Limited

Unit 7, Abbey Court, Corporation Road, Leicester LE4 5PW

Tel: +44 (0)116 2667029 • Fax: +44 (0)116 2662933

Certificate No: RSI-003

Registered Supervised Sprinkler Installer, recognised to design and install, under the supervision of an LPCB Certificated Supervising Body, sprinkler systems conforming to the *LPC Rules for Automatic Sprinkler Installations*, incorporating BS 5306: Part 2: 1990.
Supervising Company: Angus Fire Armour Limited

Fire Valve Services Limited

Broom Street, Off Huddersfield Road, Newhey, Rochdale OL16 3RY

Tel: +44 (0)1706 848599 • Fax: +44 (0)1706 843474

Certificate No: RSI-002

Registered Supervised Sprinkler Installer, recognised to design and install, under the supervision of an LPCB Certificated Supervising Body, sprinkler systems conforming to the *LPC Rules for Automatic Sprinkler Installations*, incorporating BS 5306: Part 2: 1990.
Supervising Company: Angus Fire Armour Limited

Firewatch (A Division of U.K. Fire International)

1 Winchester Drive, South West Industrial Estate, Peterlee, Co. Durham SR8 2RJ

Tel: +44 (0)191 586 0900 • Fax: +44 (0)191 586 0782

Certificate No: RSI-005

Registered Supervised Sprinkler Installer, recognised to design and install, under the supervision of an LPCB Certificated Supervising Body, sprinkler systems conforming to the *LPC Rules for Automatic Sprinkler Installations*, incorporating BS 5306: Part 2: 1990.
Supervising Company: Angus Fire Armour Limited
            Hall Fire Protection Limited

First Fire Protection Limited

Unit 7, Wycombe Industrial Mall, West End Street, High Wycombe, Buckinghamshire HP11 2QY

Tel: +44 (0)1494 522031/2 • Fax: +44 (0)1494 452752

Certificate No: RSI-022

Registered Supervised Sprinkler Installer, recognised to design and install, under the supervision of an LPCB Certificated Supervising Body, sprinkler systems conforming to the *LPC Rules for Automatic Sprinkler Installations*, incorporating BS 5306: Part 2: 1990.
Supervising Company: Angus Fire Armour Limited
            Hall Fire Protection Limited

Gordonson Fire Protection Limited

Unit 4, Almond Road, Bermondsey, London SE16 3LR

Tel: +44 (0)171 237 6707 • Fax: +44 (0)171 252 0152

Certificate No: RSI-058

Registered Supervised Sprinkler Installer, recognised to design and install, under the supervision of an LPCB Certificated Supervising Body, sprinkler systems conforming to the LPC Rules for Automatic Sprinkler Installations, incorporating BS 5306 : Part 2 : 1990.
Supervising Company: Angus Fire Armour Limited

Haden Fire Protection

95-107 Lancefield Street, Glasgow, Scotland G3 8JB

Tel: +44 (0)141 248 3701 • Fax: +44 (0)141 226 4790

Certificate No: RSI-054

Registered Supervised Sprinkler Installer, recognised to design and install, under the supervision of an LPCB Certificated Supervising Body, sprinkler systems conforming to the LPC Rules for Automatic Sprinkler Installations, incorporating BS 5306 : Part 2 : 1990.
Supervising Company: Haden Young Fire Engineering, Watford

FHC approved

 ✔

Hydra Industrial Services Limited
Bentley Centre, Stratton Road, Swindon, Wiltshire SN1 2SH

Tel: +44 (0)1793 616333 • Fax: +44 (0)1793 618333

Certificate No: RSI-048

Registered Supervised Sprinkler Installer, recognised to design and install, under the supervision of an LPCB Certificated Supervising Body, sprinkler systems conforming to the *LPC Rules for Automatic Sprinkler Installations,* incorporating BS 5306: Part 2: 1990.
Supervising Company: Actspeed Limited

J & J Design
John Street Works, John Street, Haslingden, Lancashire BB4 5QB

Tel: +44 (0)1706 223414 • Fax: +44 (0)1706 225639

Certificate No: RSI-041

Registered Supervised Sprinkler Installer, recognised to design and install, under the supervision of an LPCB Certificated Supervising Body, sprinkler systems conforming to the LPC Rules for Automatic Sprinkler Installations, incorporating BS 5306 : Part 2 : 1990.
Supervising Company: Angus Fire Armour Limited

Kean Fire Protection Services
Leveson Street, Willenhall, West Midlands WV13 1DB

Tel: +44 (0)1902 603550 • Fax: +44 (0)1902 608017

Certificate No. RSI-057
Registered Supervised Sprinkler Installer, recognised to design and install, under the supervision of an LPCB Certificated Supervising Body, sprinkler systems conforming to the LPC Rules for Automatic Sprinkler Installations, incorporating BS 5306 : Part 2 : 1990.
Supervising Companies: Angus Fire Armour Limited
Hall Fire Protection Limited

Kentallen Fire Protection Limited
Lancaster Road, Lowmoss Industrial Estate, Bishopbriggs G64 2LL

Tel: +44 (0)141 762 2540 • Fax: +44 (0)141 762 0823

Certificate No. RSI-006
Registered Supervised Sprinkler Installer, recognised to design and install, under the supervision of an LPCB Certificated Supervising Body, sprinkler systems conforming to the LPC Rules for Automatic Sprinkler Installations, incorporating BS 5306 : Part 2 : 1990.
Supervising Company: Actspeed Limited

Lecol Engineering Limited
Unit 15, Clifton Road, Huntingdon, Cambridgeshire PE18 7EJ

Tel: +44 (0)1480 436663 • Fax: +44 (0)1480 436664

Certificate No: RSI-064

Registered Supervised Sprinkler Installer, recognised to design and install, under the supervision of an LPCB Certificated Supervising Body, sprinkler systems conforming to the LPC Rules for Automatic Sprinkler Installations, incorporating BS 5306 : Part 2 : 1990.
Supervising Company: Angus Fire Armour Limited

Lostock Design Partnership

Centurion House, Centurion Way, Farrington, Leyland, Preston, Lancashire
PR5 2RE

Tel: +44 (0)1772 424252 • Fax: +44 (0)1772 424202

Certificate No: RSI-028

Registered Supervised Sprinkler Installer, recognised to design and install, under the supervision of an LPCB
Certificated Supervising Body, sprinkler systems conforming to the *LPC Rules for Automatic Sprinkler Installations*,
incorporating BS 5306: Part 2: 1990.
Supervising Company: Hall Fire Protection Limited.

Manchester Fire Protection Limited

Errwood House, 212 Moss Lane, Bramhall, Cheshire SK7 1BD

Tel: +44 (0)161 4398513 • Fax: +44 (0)161 4408930

Certificate No: RSI-004

Registered Supervised Sprinkler Installer, recognised to design and install, under the supervision of an LPCB
Certificated Supervising Body, sprinkler systems conforming to the *LPC Rules for Automatic Sprinkler Installations*,
incorporating BS5306: Part 2: 1990.
Supervising Company: Hall Fire Protection Limited.

Maximum Fire Sprinkler Systems Limited

3 Shotts Street, Queenslie Industrial Estate, Glasgow G33 4JB

Tel: +44 (0)141 774 0080 • Fax: +44 (0)141 774 7080

Certificate No: RSI-073

Registered Supervised Sprinkler Installer, recognised to design and install, under the supervision of an LPCB
Certificated Supervising Body, sprinkler systems conforming to the LPC *Rules for Automatic Sprinkler Installations*,
incorporating BS 5306 : Part 2 : 1990.
Supervising Company: Hall Fire Protection Limited.

MacLellan Engineering Services plc

Hailey Road, Thamesmead, Erith, Kent DA18 4AH

Tel: +44 (0)181 310 5000 • Fax: +44 (0)181 311 4618

Certificate No: RSI-059

Registered Supervised Sprinkler Installer, recognised to design and install, under the supervision of an LPCB
Certificated Supervising Body, sprinkler systems conforming to the LPC Rules for Automatic Sprinkler Installations,
incorporating BS 5306 : Part 2 : 1990.
Supervising Company: Angus Fire Armour Limited.

Parkinson Fire Engineering Services Limited

Leek Road, Stoke-on-Trent, Staffordshire ST4 2EJ

Tel: +44 (0)1782 289707 • Fax: +44 (0)1782 202 394

Certificate No: RSI-047

Registered Supervised Sprinkler Installer, recognised to design and install, under the supervision of an LPCB
Certificated Supervising Body, sprinkler systems conforming to the *LPC Rules for Automatic Sprinkler Installations*,
incorporating BS 5306: Part 2: 1990.
Supervising Company: Angus Fire Armour Limited.

Peterlee Fire Company Limited
1 Winchester Drive, S.W. Industrial Estate, Peterlee, Co. Durham Sr8 2RJ

Tel: +44 (0)191 586 0900 • Fax: +44 (0)191 586 0782

Certificate No: RSI-065

Registered Supervised Sprinkler Installer, recognised to design and install, under the supervision of an LPCB Certificated Supervising Body, sprinkler systems conforming to the *LPC Rules for Automatic Sprinkler Installations*, incorporating BS5306: Part 2: 1990.
Supervising Company: Angus Fire Armour Limited.

Protec Camerfield Limited
5, Churchill Way, Nelson, Lancashire BB9 6RT

Tel: +44 (0)1282 717474 • Fax: +44 (0)1282 717475

Certificate No: RSI-050

Registered to Supervised Sprinkler Installer, recognised to design and install, under the supervision of an LPCB Certificated Supervising Body, sprinkler systems conforming to the *LPC Rules for Automatic Sprinkler Installations*, incorporating BS 5306: Part 2: 1990.
Supervising Company: Angus Fire Armour Limited.

Red Rose Mechanical Services
Unit 22, South Ribble Enterprise Park, South Ribble Industrial Estate, Capitol Way, Walton-Le-Dale, Preston PR5 4AQ

Tel:+44 (0)1772 886646 • Fax: +44 (0)1772 881702

Certificate No. RSI-031
Registered Supervised Sprinkler Installer, recognised to design and install, under the supervision of an LPCB Certificated Supervising Body, sprinkler systems conforming to the LPC Rules for Automatic Sprinkler Installations, incorporating BS 5306 : Part 2 : 1990.
Supervising Company: Angus Fire Armour Limited

R.G.T. Fire Protection
95 Beaumont Road, Bournville, Birmingham B30 2EB

Tel: +44 (0)121 433 5175 • Fax: +44 (0)121 486 1847

Certificate No: RSI-069
Registered Supervised Sprinkler Installer, recognised to design and install, under the supervision of and LPCB Certificated Supervising Body, sprinkler systems conforming to the LPC Rules for Automatic Sprinkler Installations, incorporating BS 5306 : Part 2 : 1990.
Supervising Companies: Angus Fire Armour Limited

Richmond Fire Engineers Limited
30 Firby Road, Gallowfields Trading Estate,Richmond, North Yorkshire DL10 4ST

Tel: +44 (0)1748 825612 • Fax: +44 (0)1748 825935

Certificate No: RSI-053

Registered Supervised Sprinkler Installer, recognised to design and install, under the supervision ofan LPCB Certificated Supervising Body, sprinkler systems conforming to the LPC Rules for Automatic Sprinkler Installations, incorporating BS 5306 : Part 2 : 1990.
Supervising Company: Angus Fire Armour Limited.

RMD Fire Control Limited
Ramsden House, 23 Wrotham Road, Gravesend, Kent DA11 0PA

Tel: +44 (0)1474 564411 • Fax: +44 (0)1474 564472

Certificate No: RSI-020

Registered Supervised Sprinkler Installer, recognised to design and install, under the supervision of an LPCB
Certificated Supervising Body, sprinkler systems conforming to the *LPC Rules for Automatic Sprinkler Installations,*
incorporating BS 5306: Part 2: 1990.
Supervising Company: Angus Fire Armour Limited

RSM Refko Installations Limited
Lambs Lane, Rainham, Essex RM13 9XL

Tel: +44 (0)1708 520236 • Fax: +44 (0)1708 521379

Certificate No: RSI-030

Registered Supervised Sprinkler Installer, recognised to design and install, under the supervision of an LPCB
Certificated Supervising Body, sprinkler systems conforming to the LPC Rules for Automatic Sprinkler Installations,
incorporating BS 5306 : Part 2 : 1990.
Supervising Company: Angus Fire Armour Limited.

Sentinel Sprinklers Limited ✔
6 Newport Road, Trethomas, Gwent NP1 8BY

Tel: +44 (0)1222 865511 • Fax: +44 (0)1222 862054

Certificate No: RSI-013

Registered Supervised Sprinkler Installer, recognised to design and install, under the supervision of an LPCB
Certificated Supervising Body, sprinkler systems conforming to the *LPC Rules for Automatic Sprinker Installations,*
incorporating BS 5306: Part 2: 1990.
Supervising Company: Angus Fire Armour Limited

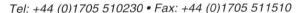

Solent Fire Protection Services Limited ✔
Unit 22, Westfield Industrial Estate, Westfield Road, Gosport, Hampshire PO12 3RX

Tel: +44 (0)1705 510230 • Fax: +44 (0)1705 511510

Certificate No: RSI-051

Registered Supervised Sprinkler Installer, recognised to design and install, under the supervision of an LPCB
Certificated Supervising Body, sprinkler systems conforming to the *LPC Rules for Automatic Sprinkler Installations,*
incorporating BS5306: Part 2: 1990.
Supervising Company: Angus Fire Armour Limited

Standguard Sprinkler Systems Limited
114 Fearns Avenue, Bradwell, Newcastle-Under-Lyme, Staffs ST5 8LX

Tel: +44 (0)1782 660230 • Fax: +44 (0)1782 712637

Certificate No: RSI-074

Registered Supervised Sprinkler Installer, recognised to design and install, under the supervision of an LPCB
Certificated Supervising Body, sprinkler systems conforming to the LPC Rules for Automatic Sprinkler Installations,
incorporating BS 5306 : Part 2 : 1990.
Supervising Company: Angus Fire Armour Limited.

5 SECTION 1B:
LPS 1048 REGISTERED SUPERVISED SPRINKLER INSTALLERS

FHC approved

Taylor Robinson Limited
Fire Protection House, Woolley Colliery Road, Darton, Barnsley, Yorkshire S75 5JA

Tel: +44 (0)1226 386911/2/3 • Fax: +44 (0)1226 388206

Certificate No: RSI-001

Registered Supervised Sprinkler Installer, recognised to design and install, under the supervision of an LPCB
Certificated Supervising Body, sprinkler systems conforming to the *LPC Rules for Automatic Sprinkler Installations*,
incorporating BS 5306: Part 2: 1990.
Supervising Company: Angus Fire Armour Limited
                                          Actspeed Limited

Thameside Fire Protection Co. Limited
Unit 7, Swinborne Court, Burnt Mills Industrial Estate, Basildon, Essex SS13 1QA

Tel: +44 (0)1268 591059 • Fax: +44 (0)1268 590974

Certificate No: RSI-046

Registered Supervised Sprinkler Installer, recognised to design and install, under the supervision of an LPCB
Certificated Supervising Body, sprinkler systems conforming to the *LPC Rules for Automatic Sprinkler Installations*,
incorporating BS 5306: Part 2: 1990.
Supervising Company: Hall Fire Protection Limited

Wormald Ansul (UK) Limited
Wormald Fire Systems
Hertburn Industrial Estate, Washington, Tyne & Wear NE37 2SF

Tel: +44 (0)191 416 2651 • Fax: +44 (0)191 415 3084

Certificate No: RSI-024

Registered Supervised Sprinkler Installer, recognised to design and install, under the supervision of an LPCB
Certificated Supervising Body, sprinkler systems conforming to the *LPC Rules for Automatic Sprinkler Installations*,
incorporating BS 5306: Part 2: 1990.
Supervising Company: Wormald Ansul (UK) Limited, Wormald Fire Systems - Manchester, Smethwick and Slough

Writech Industrial Services Limited
Newbrook, Mullingar, Co. Westmeath, Ireland

Tel: +353 44 49857 • Fax: +353 44 49858

Certificate No: RSI-070

Registered Supervised Sprinkler Installer, recognised to design and install, under the supervision of an LPCB
Certificated Supervising Body, sprinkler systems conforming to the *LPC Rules for Automatic Sprinkler Installations*,
incorporating BS 5306: Part 2: 1990.
Supervising Company: Actspeed Limited

Hall Fire Protection Limited

186 Moorside Road, Swinton, Manchester M27 9HA

Tel: +44 (0)161 793 4822 • Fax: +44 (0)161 794 4950

Certificate No: CSM-001

Certificated Sprinkler Servicing and Maintenance Firm, authorised to issue LPS 1050 Certificates of Servicing and Inspection for sprinkler systems serviced and maintained in accordance with BS 5306: Part 2: 1990.

LPS 1050 Certificated Supervising Body, authorised to supervise LPS 1050 Registered Servicing Firms

Preussag Fire Protection Limited

Field Way, Greenford, Middlesex UB6 8UZ

Tel: +44 (0)181 832 2000 • Fax: +44 (0)181 832 2200

Certificate No: CSM-002

Certificated Sprinkler Servicing and Maintenance Firm, authorised to issue LPS 1050 Certificates of Servicing and Inspection for sprinkler systems serviced and maintained in accordance with BS 5306: Part 2: 1990.

5 SECTION 3:
3.1: ALARM AND DRY PIPE VALVES

It is recommended that all alarm valves, dry-pipe valves and back-pressure valves be the subject of a maintenance contract, satisfying the requirements of the LPC *Rules for Automatic Sprinkler Installations.*

The nominal orifice size of valves is given in millimetres.

AFAC Inc.
West 2nd Avenue, Ranson, West Virginia 25438, USA

Tel: +1 304 728 9000 • Fax: +1 304 728 8481

Automatic Dry Pipe Valve	(Model 39)	100mm 150mm
Automatic Alarm Valve	(Model 353)	100mm 150mm 200mm
Automatic Alarm Valve	(Model 363)	100mm 150mm

Angus Fire Armour Limited
Kastanievej 15, DK-5620, Glamsbjerg, Denmark

UK Sales Enquiries: Tel: +44 (0)1844 214545 • Fax: +44 (0)1844 213511

Angus Alarm Valve [1]	Model (C) vertical (cast-iron)	80, 100, 150, 200 mm
Angus Alarm Valve	Model (C1) vertical (cast-iron)	80, 100, 150, 200 mm
Angus Alarm Valve[1]	(Model C) vertical (leaded gunmetal)	100mm
Angus Alarm Valve[1]	(Model C) vertical (H.T. aluminium bronze)	100mm
Angus Dry Pipe Valve[2]	(Model D) vertical (cast iron)	100mm 150mm
Angus Dry Pipe Valve	(Model 39)	100, 150mm

[1] Maximum working pressures: 20 bar.
[2] With flanged inlet and flanged or grooved outlet

Central Sprinkler Company
451 N. Cannon Avenue, Lansdale, Pennsylvania 19446, USA

Tel: +1 215 362 0700 • Fax: +1 215 362 5385

Central Alarm Valve	(Model F)	100mm 150mm 200mm
Central Dry Pipe Valve	(Model D)	50mm 63mm 75mm 100mm 150mm
Central Dry Pipe Valve[1]	(Model H)	100mm 150mm
Central Dry Pipe Valve	(Model AF)	100mm 150mm
Central Dry Pipe Valve	(Model AG)	100mm 150mm

[1] Maximum working pressure: 12 bar.

Globe Fire Sprinkler Corporation
4077 Air Park Drive, P.O. Box 796, Standish, Michigan 48658, USA

Tel: +1 517 846 4583 • Fax: +1 517 846 9231

Globe Alarm Valve vertical. (Model H1), 100mm[1], 150mm[1].
Globe Alarm Valve[1] vertical. (Model H3), 100mm[2], 150mm[2].
[1] Maximum working pressure 20 bar.
[2] Maximum working pressure 12 bar.

Globe Sprinklers Europa S.A.
Avda. de las Flores, 13 Parque Empresarial "El Molino", 28970 Humanes de Madrid (Madrid), Spain

Tel: +34 1 606 3711 • Fax: +34 1 690 9561

Globe Alarm Valve   Model H-1   100mm[1],150mm[1]

[1]. Maximum working pressure 16 bar

Grinnell Corporation

PO Box 128, Highway 70, Cleveland (Rowan County), North Carolina 27013, USA

Tel: +1 704 278 2221 • Fax: +1 704 278 9617
Sales Enquiries: Tel: +31 5 328 3434 • Fax: +31 5 328 3377

Gem Dry Pipe Valve	(Model A2)	50mm
Gem Dry Pipe Valve[1,2]	(Model F302)	100mm 150mm
Gem Dry Pipe Valve[1,2]	(Model F3021)	100mm 150mm
Gem Alarm Valve[3]	(Model F200)	100mm 150mm 200mm
Gem Alarm Valve[3]	(Model F2001)	100mm 150mm 200mm
Gem Dry Pipe Valve	(Model F3061)	50mm, 75mm

[1] Only approved when trimmed in accordance with Grinnell bulletins, Reference TD107.
[2] Installations containing models F302 and F3021 shall be maintained under servicing contracts with the manufacturer, or
    LPS 1048 and LPS 1050 firms experienced in the installation and servicing of this valve.
[3] Maximum working pressure of 12 bar.

Grinnell Manufacturing (UK) Limited

Stockport Trading Estate, Yew Street, Stockport SK4 2JW

Tel: +44 (0)161 477 1886 • Fax: +44 (0)161 477 6729

B.W. Alarm Valve[1,2]	vertical	(gunmetal)	Model No. FE 2394	50/80mm
B.W. Alarm Valve [1]	vertical	(gunmetal)	Model No. FE 2217	100mm
B.W. Alarm Valve [1]	vertical	(stainless steel)	Model No. FE 2640	100mm
B.W. Alarm Valve [1]	horizontal	(gunmetal)	Model No. FE 2375	100mm
B.W. Alarm Valve [1]	horizontal	(stainless steel)	Model No. FE 2646	100mm
B.W. Alarm Valve [1]	vertical	(aluminium bronze)	Model No. FE 2217	100mm
B.W. Alarm Valve [3]	vertical	(gunmetal)	Model No. FE 3557	100mm
B.W. Alarm Valve [1]	vertical	(gunmetal)	Model No. FE 2188	150mm
B.W. Alarm Valve [1]	vertical	(stainless steel)	Model No. FE 2643	150mm
B.W. Alarm Valve [1]	horizontal	(gunmetal)	Model No. FE 2354	150mm
B.W. Alarm Valve [1]	horizontal	(stainless steel)	Model No. FE 2649	150mm
B.W. Alarm Valve [3]	vertical	(aluminium bronze)	Model No. FE 3262	150mm
B.W. Alarm Valve [1]	vertical	(bronze)	Model No. FE 2360	200mm
B.W. Alarm Valve [1]	horizontal	(bronze)	Model No. FE 2685	200mm
Wet Pipe Alarm Valves[4]			Model L/W	80,100,150, 200mm
B.W. Alternate Valve [1,5]	vertical	(gunmetal)	Model No. FE 2349	100mm
B.W. Alternate Valve [1,5]	vertical	(stainless steel)	Model No. FE 2642	100mm
B.W. Alternate Valve [1,5]	vertical	(gunmetal)	Model No. FE 2152	150mm
B.W. Alternate Valve [1,5]	vertical	(stainless steel)	Model No. FE 2645	150mm
B.W. Alternate Valve [1,5]	vertical	(bronze)	Model No. FE 2380	200mm
Wormald Alarm Valve	vertical & horizontal	(cast iron)	Model NV	100mm
Wormald Alarm Valve	vertical & horizontal	(cast iron)	Model NV	150mm
Wormald Alarm Valve	vertical & horizontal	(cast iron)	Model NV	200mm

[1] Maximum working pressure : 16 bar.
[2] Can be used with either 50mm or 80mm nominal bore pipe.
[3] Maximum working pressure : 25 bar.
[4] With flanged inlet and flanged or grooved outlet.
[5] Installations containing the BW alternate valve shall be maintained under servicing contracts with the manufacturer, or LPS
1048 and LPS 1050 firms experienced in the installation and servicing of this valve.

5 SECTION 3:

3.1: ALARM AND DRY PIPE VALVES

GW Sprinkler A/S
Kastanievej 15, DK-5620 Glamsbjerg, Denmark

Tel: +45 64 72 2055 • Fax: +45 64 72 2255

G. W. Alarm Valve	(Model C1)	80, 100, 150, 200mm
G. W. Dry Pipe Valve[1]	(Model D) vertical (cast iron)	100, 150mm
GW Horizontal Alarm Valve	(Model C1-H)	80mm[2]
GW Dry Pipe Valve	(Model 39)	100, 150mm

[1] With flanged inlet and flanged or grooved outlet.
[2] Maximum working pressure 20 bar.

Minimax GmbH
Industriestrasse 10/12, D-23840 Bad Oldesloe, P O Box 1260, Germany

Tel: +49 4531 8030 • Fax: +49 4531 803248
UK Sales Enquiries: Tel: +44 (0)181 832 2000 • Fax: +44 (0)181 832 2200

MX Alarm Valve Type NAV (Fullway pattern)	100mm 150mm 200mm
MX Dry Pipe Alarm Valve Type TAV	80mm 100mm
MX Alarm Valve Type AV	150mm
(Suitable for wet and dry pipe systems)	
TAV Dry Pipe Alarm Valve Type TMX	80mm, 100mm, 150mm
Wet Alarm Valve Type NMX[1]vertical (cast iron)	80mm, 100mm, 150mm, 200mm

[1]With flanged inlet and flanged or grooved outlet

The Reliable Automatic Sprinkler Co. Inc.
525 North MacQuesten Parkway, Mount Vernon, New York, 10552-2600 USA

Tel: +1 914 668 3470 • Fax: +1 914 668 2936
UK Sales Enquiries: Tel: +44 (0)11372 728899 • Fax: +44 (0)1372 724461

Reliable Alarm Valve (Model E) with flanged inlet and outlet	65mm 100mm 150mm 200mm
Reliable Dry Pipe Valve (Model D) with flanged inlet and flanged or grooved outlet	100mm 150m
Reliable Dry Pipe Valve (Model A) with threaded inlet and outlet	65mm

---

Spraysafe Automatic Sprinklers Limited
Corringham Road Industrial Estate, Gainsborough, Lincolnshire DN 21 1QB

Tel: +44 (0)1427 615401 • Fax: +44 (0)1427 610433

Spraysafe Wet Alarm Valve	(Model F)	100mm[1], 150mm[1] 200mm
Spraysafe Dry Pipe Valve	(Model D)	50mm 63mm 75mm
Spraysafe Dry Pipe Valve	(Model AF)	100mm 150mm
Spraysafe Dry Pipe Valve	(Model AG)	100mm 150mm

[1]Maximum working pressure 20 bar

Star Sprinkler Inc.
PO Box 128, Highway 70, Cleveland (Rowan County), North Carolina 27013, USA

Tel: +1 414 570 5000 • Fax: +1 414 570 5010

Star Dry Pipe Valve	(Model A)	50mm 75mm 100mm 150mm
Star Alarm Valve	(Model E)	200mm
Star Alarm Valve	(Model F)	100mm 150mm
Star Alarm Valve[1]	(Model S300)	100mm, 150mm, 200mm
Star Alarm Valve[1]	(Model S3001)	100mm, 150mm, 200mm

[1] Maximum working pressure of 12 bar

Total Walther GmbH
Feuerschutz und Sicherheit, Waltherstr 51, 51069 Köln-Dellbruck, Germany

Tel: +49 221 67 85 631 • Fax: +49 221 67 85 727

Walther Alarm Valve (Type NV)		100mm 150mm 200mm

The Viking Corporation
210 N Industrial Park, Hastings, Michigan 49058, USA

Tel: +1 616 945 9501 • Fax: +1 616 945 9599
UK Sales Enquiries: Tel: +44 (0)1427 875999 • Fax: +44 (0)1427 875998

Viking Alarm Valve	(Model F1)	65mm 100mm 150mm 200mm
Viking Cast Iron Dry Pipe Valve[1]	(Model E-2)	75mm 100mm
Viking Cast Iron Dry Pipe Valve[1]	(Model E-1)	150mm
Viking Alarm Valve	(Model H-2)	75mm, 100mm, 150mm, 200mm
		(Maximum working pressure 12 bar)
Viking Wet Alarm Valve	(Model J-1)	75mm,100mm,150mm,200mm

[1] Approved with flanged inlet and flanged or grooved outlet.

Angus Fire Armour Limited

Kastanievej 15, DK-5620, Glamsbjerg, Denmark

UK Sales Enquiries: Tel: +44 (0)1844 214545 • Fax: +44 (0)1844 213511

Type	Size (mm)	Valve body Model	Material	Alarm motor and gong	Maximum working pressure (bar)	Attitude
Crusader W A	80	C1	Cast Iron	D, E	12.5	Vertical
Crusader WA	100	C1	Cast Iron	D, E	12.5	Vertical
Crusader WA	150	C1	Cast Iron	D, E	12.5	Vertical
Crusader WA	200	C1	Cast Iron	D, E	12.5	Vertical

The Angus Crusader wet alarm valves are supplied with Angus data sheet 020, which includes installation, commissioning and operating instructions.

It is recommended that all alarm valves and back-pressure valves be the subject of a maintenance contract, satisfying the requirements of the *LPC Rules for automatic sprinkler installations.*

AFAC Inc.
West 2nd Avenue, Ranson, West Virginia 25438, USA

Tel: +1 304 728 9000 • Fax: +1 304 728 8481

Automatic Accelerator (Model 1)

Angus Fire Armour Limited
Kastanievej 15, DK-5620, Glamsbjerg, Denmark

UK Sales Enquiries: Tel: +44 (0)1844 214545 • Fax: +44 (0)1844 213511

Angus Fire Armour
Angus Accelerator (Model BI)

Central Sprinkler Company
451 N. Cannon Avenue, Lansdale Pennsylvania 19446, USA

Tel: +1 215 362 0700 • Fax: +1 215 362 5385

Central Model A Accelerator

Grinnell Corporation
PO Box 128, Highway 70, Cleveland (Rowan County), North Carolina 27013, USA

Tel: +1 704 278 2221 • Fax: +1 704 278 9617
Sales Enquiries: Tel: +31 53 28 34 34 • Fax: +31 53 28 33 77

Gem Accelerator (Model F311)

Grinnell Manufacturing (UK) Limited
Stockport Trading Estate, Yew Street, Stockport SK4 2JW

Tel: +44 (0)161 477 1886 • Fax: +44 (0)161 477 6729

Model No. 936 Accelerator

Minimax GmbH
Industriestrasse 10/12, D-23840 Bad Oldesloe, Germany

Tel: +49 4531 803 0 • Fax: +49 4531 803 248

MX Accelerator (Model 12296693) (previously listed Model 5113).

The Reliable Automatic Sprinkler Co. Inc.
525 North MacQuesten Parkway, Mount Vernon, New York, 10552-2600, USA

Tel: +1 914 668 3470 • Fax: +1 914 668 2936
UK Sales Enquiries: Tel: +44 (0)1372 728899 • Fax: +44 (0)1372 724461

Reliable Accelerator (Model B1).

Spraysafe Automatic Sprinklers Limited
Corringham Road Industrial Estate, Gainsborough, Lincolnshire DN21 1QB

Tel: +44 (0)1427 615401 • Fax: +44 (0)1427 610433

Spraysafe Model A Accelerator

Star Sprinkler Inc.
Highway 70, PO Box 128, Cleveland (Rowan County), North Carolina 27013, USA

Tel: +1 414 570 5000 • Fax: +1 414 570 5010

Star (Model BB) Exhauster
Star Accelerator (Model S430)

The Viking Corporation
210 N Industrial Park Road, Hastings, Michigan 49058, USA

Tel: +1 616 945 9501 • Fax: +1 616 945 9599
UK Sales Enquiries
Tel: +44 (0)1427 875999 • Fax: +44 (0)1427 875998

Viking Accelerator (Model B)
Viking Accelerator (Model D-1)

It is recommended that all alarm valves and back-pressure valves be the subject of a maintenance contract, satisfying the requirements of the *LPC Rules for automatic sprinkler installations.*

AFAC Inc.
West 2nd Avenue, Ranson, West Virginia 25438, USA

Tel: +1 304 728 9000 • Fax: +1 304 728 8481

Automatic Deluge Valve	(Model C)	65mm 150mm

Angus Fire Armour Limited
Kastanievej 15, DK-5620, Glamsbjerg, Denmark

UK Sales Enquiries: Tel: +44 (0)184 421 4545 • Fax: +44 (0)184 421 3511

Angus Deluge Valve Mk.III[1]	(aluminium bronze)	100mm 150mm 200mm
Angus Deluge Valve Mk.III[2]	(cast iron)	100mm 150mm 200mm

[1] Maximum working pressure: 20 bar
[2] Maximum working pressure: 16 bar

Grinnell Corporation
PO Box 128, Highway 70, Cleveland (Rowan County), North Carolina 27013, USA

Tel: +1 704 278 2221 • Fax: +1 704 278 9617
Sales Enquiries: Tel: +31 53 428 3434 • Fax: +31 53 428 3377

Gem Deluge Valve	(Model A-4)	100mm 150mm
Gem Deluge Valve	(Model B)	50mm
Gem Dry Pilot Actuator	(Model B1)	
Gem Deluge Valve	Model F445	65mm
Gem Deluge Valve	Model F470	100mm, 150mm

Grinnell Manufacturing (UK) Limited
Stockport Trading Estate, Yew Street, Stockport SK4 2JW

Tel: +44 (0)161 477 1886 • Fax: +44 (0)161 477 6729

B.W. Deluge Valve[1] vertical	(gunmetal)	Model No. FE 2408	80mm
B.W. Deluge Valve[1] vertical	(gunmetal)	Model No. FE 2333	100mm
B.W. Deluge Valve[1] vertical	(stainless steel)	Model No. FE 2641	100mm
B.W. Deluge Valve[1] horizontal	(gunmetal)	Model No. FE 2366	100mm
B.W. Deluge Valve[1] horizontal	(stainless steel)	Model No. FE 2647	100mm
B.W. Deluge Valve[1] vertical	(gunmetal)	Model No. FE 2326	150mm
B.W. Deluge Valve[1] vertical	(stainless steel)	Model No. FE 2644	150mm
B.W. Deluge Valve[1]horizontal	(gunmetal)	Model No. FE 2353	150mm
B.W. Deluge Valve[1] horizontal	(stainless steel)	Model No. FE 2650	150mm
B.W. Deluge Valve[1] vertical	(bronze)	Model No. FE 2370	200mm
B.W. Deluge Valve[1] horizontal	(bronze)	Model No. FE 2687	200mm

[1]Maximum working pressure : 16 bar

G W Sprinkler A/S
Kastanievej 15, DK-5620, Glamsbjerg, Denmark

Tel: +45 64 72 2055 • Fax: +45 64 72 2255

GW Deluge Valve (Model B) 100mm, 150mm.

Minimax GmbH
Industriestrasse 10/12, D-23840 Bad Oldesloe, Germany

Tel: +49 4531 803 0 • Fax: +49 4531 809 248

Minimax Deluge Valve	50mm 80mm 100mm 150mm

The Reliable Automatic Sprinkler Co. Inc.
525 North MacQuesten Parkway, Mount Vernon, New York, 10552-2600 USA

Tel: +1 914 668 3470 • Fax: +1 914 668 2936
UK Sales Enquiries: Tel: +44 (0)1372 728899 • Fax: +44 (0)1372 724461

Reliable Deluge Valve	(Model B)	100mm 150mm
Reliable Deluge Valve	(Model A) (screwed inlet and outlet)	65mm
Reliable Dry Pilot Line Actuator		

Star Sprinkler Inc.
Highway 70, PO Box 128, Cleveland (Rowan County), North Carolina 27013, USA

Tel: +1 414 570 5000 • Fax: +1 414 570 5010

Star Deluge Valve	(Model G)	80mm, 150mm
Star Deluge Valve	(Model B)	50mm

Total Walther GmbH
Feuerschutz und Sicherheit, Waltherstr 51, 51069 Kôln-Dellbruck, Germany.

Tel: +49 221 67 85 631 • Fax: +49 221 67 85 727

Walther Deluge Valve	80mm 100mm 150mm

The Viking Corporation
210 N Industrial Park Road, Hastings, Michigan 49058, USA

Tel: +1 616 945 9501 • Fax: +1 616 945 9599
UK Sales Enquiries: Tel: +44 (0)1427 875999 • Fax: +44 (0)1427 875998

Viking Deluge Valve	(Model E-1) Screwed Inlet and Outlet	50mm
Viking Deluge Valve	(Model E-1) Flanged Inlet and Outlet	80mm 100mm 150mm
Viking Deluge Valve	(Model E-1) Flanged Inlet and Grooved Outlet	80mm 100mm 150mm

SECTION 6:
DIRECT READING FLOW METERS

The following devices are approved for use under the *LPC Rules for automatic sprinkler installations.*

Grinnell Manufacturing (UK) Limited
Stockport Trading Estate, Yew Street, Stockport, SK4 2JW

Tel: +44 (0)161 477 1886 • Fax: +44 (0)161 477 6729

Hazard class	Pipe size mm	Flow range dm³/min	Orifice size mm	Ref. no.
OHI	80	375 - 540	26.10	F2/1
OHII	80	725 - 1000	39.25	F2/2
OHIII	80	1100 - 1350	43.70	F2/3
OHIIIS	80	1800 - 2100	55.20	F2/4

These measuring devices are designed to be fitted directly on the upstream side, to straight pipes or into elbow fittings or globe valves or their hydraulic equivalent. The downstream flow loss may be zero in all but the case of the OHI measuring device which requires a minimum loss of 0.02 bar at a flow of 375 dm³/min. The maximum downstream flow loss shall not exceed 0.3 bar for any of the measuring devices with the following respective flow rates:
OHI 540dm³/min,   OHII 1000dm³/min,   OHIII 1350dm³/min,   OHIIIS 2100dm³/min

Platon Instrumentation Limited
Platon Park, Viables, Basingstoke, Hampshire RG22 4BS

Tel: +44 (0)1256 470456 • Fax: +44 (0)1256 363345

Type	Pipe size mm	Flow range dm³/min	Ref. no.
Shunt Gapmeters	80	300 - 1500	F1/1
	100	500 - 2500	F1/2
	125	700 - 3500	F1/3
	150	1000 - 5000	F1/4

These meters have been approved for installation with the gauge glass in the vertical position. There must be at least five diameters of straight pipe upstream and downstream of the orifice plate. The isolating valves on the pressure tappings must be kept closed when the meter is not being used. A spare gauge glass must be provided and located in the cabinet containing the stock of replacement sprinkers.

Type	Nominal pipe size to		Flow range	Ref. no.
	BS 1387 mm	BS 3600 mm	dm³/min	
Type SGUV Mk 2[1]	50		150 - 750	F1/5
	80		300 - 1500	F1/6
	100		500 - 2500	F1/7
	150		1000 - 5000	F1/8
		200	2500 -12500	F1/9

[1] Approved to the *FOC Requirements and Testing Methods for Direct Reading Flow Meters:* 1982, and to be installed in accordance with G. A. Platon Document OM 1001.

† Indicates glass bulb type; all others are solder types.
* Indicates that a corrosion-resistant coating for the body and yoke has been approved under the *LPC Rules for automatic sprinkler installations.*

Angus Fire Armour Limited
Kastanievej 15, DK-5620, Glamsbjerg, Denmark

UK Sales Enquiries: Tel: +44 (0)1844 214545 • Fax: +44 (0)1844 213511

† Angus Multiple Control Model B (Double Outlet)	40mm
† Angus Multiple Control Model B (Double Outlet)	50mm
† Angus Multiple Control Model B3 (Single Outlet)[1]	20mm

[1] Tested to the *FOC Requirements and Testing Methods for Multiple Controls*, April 1984 and approved with plain and chromium finish

Grinnell Corporation
PO Box 128, Highway 70, Cleveland (Rowan County), North Carolina 27013, USA

Tel: +1 704 278 2221 • Fax: +1 704 278 9617
Sales Enquiries: Tel: +31 53 428 3434 • Fax: +31 53 428 3377

Gem Multiple Control Type F430 (Single Outlet)	25mm 40mm

Grinnell Manufacturing (UK) Limited
Stockport Trading Estate, Yew Street, Stockport SK4 2JW

Tel: +44 (0)161 477 1886 • Fax: +44 (0)161 477 6729

†* Ansul Multiple Control Quartzoid (Double Outlet)	20mm
†* Ansul Multiple Control Quartzoid (Single Outlet)	20mm
†* Ansul 'In rack' Multiple Control Quartzoid (Quadruple Outlet)	25mm

Minimax GmbH
Industriestrasse 10/12, D-23840 Bad Oldesloe, Germany

Tel: +49 4531 803 0 • Fax: +49 4531 803 248

† Minimax Multiple Control (Single Outlet)	25mm 37mm 50mm

The use of proprietary pipe couplings in LPC installations is limited to where ambient temperatures do not exceed 70° C.

The maximum working pressure (M.W.P.) of listed couplings and fittings is based on proof pressure testing at 4 x M.W.P. No approval pressure testing is carried out above 80 bar.

Listed pipe coupling and fitting approvals are only valid when the equipment is installed in accordance with the manufacturer's latest installation instructions.

The pipe outside diameter(s) shall be permanently marked on all approved couplings and fittings.

Fittings are only approved for use in combination with approved couplings manufactured by the same company.

KEY

R: rolled groove, C: cut groove, T: threaded, G: grooved, S: silicone

---

Grinnell Corporation

1141 Lancaster Avenue, Columbia, Pennsylvania 17512, USA

Tel: +1 717 684 4400 • Fax: +1 717 684 2131
Sales Enquiries: Tel: +31 53 428 3434 • Fax: +31 53 428 3377

Pipe Couplings (GRUVLOK)

Type	Nominal size (mm)	Pipe outside diameter (mm)	Pipe ends	Gasket type	Maximum working pressure (bar)
Style 7000	25	33.7	R&C	EPDM	20
	32	42.4	R&C	EPDM	20
	40	48.3	R&C	EPDM	20
	50	60.3	R&C	EPDM	20
	80	88.9	R&C	EPDM	20
	100	114.3	R&C	EPDM	20
	150	168.3	R&C	EPDM	20
	200	219.1	R&C	EPDM	20
Fig 7000	65	76.1	R&C	EDPM	20
		108[1]	R&C	EPDM	20
		133[1]	R&C	EPDM	20
		159[1]	R&C	EPDM	20
	150	165.1	R&C	EPDM	20
Style 7400	25	33.7	R&C	EPDM	20
Rigidlite	32	42.4	R&C	EPDM	20
	40	48.3	R&C	EPDM	20
	50	60.3	R&C	EPDM	20
	65	76.1	R&C	EPDM	20
	80	88.9	R&C	EPDM	20
	100	114.3	R&C	EPDM	20
	125	139.7	R&C	EPDM	20
	150	165.1	R&C	EPDM	20
	150	168.3	R&C	EPDM	20
	200	219.1	R&C	EPDM	20
Style 7401	200	219.1	R&C	EPDM	20
Rigidlok					
Mechanical Tees					
Fig 7045	50 x 15	60.3 x 21.3	T	EPDM	20
	50 x 20	60.3 x 26.7	T	EPDM	20
	50 x 25	60.3 x 33.7	T	EPDM	20

Continued

159

SECTION 8:
8.1 PIPE COUPLINGS AND FITTINGS

Grinnell Corporation (continued)

Type	Nominal size (mm)	Pipe outside diameter (mm)	Pipe ends	Gasket type	Maximum working pressure (bar)
Fig 7045	50 x 32	60.3 x 42.4	T	EPDM	20
	50 x 40	60.3 x 48.3	T	EPDM	20
	65 x 15	76.1 x 21.3	T	EPDM	20
	65 x 20	76.1 x 26.7	T	EPDM	20
	65 x 25	76.1 x 33.7	T	EPDM	20
	65 x 32	76.1 x 42.4	T	EPDM	20
	65 x 40	76.1 x 48.3	T	EPDM	20
	80 x 15	88.9 x 21.3	T	EPDM	20
	80 x 20	88.9 x 26.7	T	EPDM	20
	80 x 25	88.9 x 33.7	T	EPDM	20
	80 x 32	88.9 x 42.4	T	EPDM	20
	80 x 40	88.9 x 48.3	T	EPDM	20
	80 x 50	88.9 x 60.3	T	EPDM	20
	100 x 15	114.3 x 21.3	T	EPDM	20
	100 x 20	114.3 x 26.7	T	EPDM	20
	100 x 25	114.3 x 33.7	T	EPDM	20
	100 x 32	114.3 x 42.4	T	EPDM	20
	100 x 40	114.3 x 48.3	T	EPDM	20
	100 x 50	114.3 x 60.3	T	EPDM	20
	125 x 32	139.7 x 42.4	T	EPDM	20
	125 x 40	139.7 x 48.3	T	EPDM	20
	125 x 50	139.7 x 60.3	T	EPDM	20
	150 x 32	165.1 x 42.4	T	EPDM	20
	150 x 40	165.1 x 48.3	T	EPDM	20
	150 x 50	165.1 x 60.3	T	EPDM	20
	150 x 32	168.3 x 42.4	T	EPDM	20
	150 x 40	168.3 x 48.3	T	EPDM	20
	150 x 50	168.3 x 60.3	T	EPDM	20
Fig 7046	65 x 32	76.1 x 42.4	G	EPDM	20
	65 x 40	76.1 x 48.3	G	EPDM	20
	80 x 32	88.9 x 42.4	G	EPDM	20
	80 x 40	88.9 x 48.3	G	EPDM	20
	80 x 50	88.9 x 60.3	G	EPDM	20
	100 x 32	114.3 x 42.4	G	EPDM	20
	100 x 40	114.3 x 48.3	G	EPDM	20
	100 x 50	114.3 x 60.3	G	EPDM	20
	100 x 65	114.3 x 76.1	G	EPDM	20
	100 x 80	114.3 x 88.9	G	EPDM	20
	125 x 32	139.7 x 48.3	G	EPDM	20
	125 x 50	139.7 x 60.3	G	EPDM	20
	125 x 80	139.7 x 88.9	G	EPDM	20
	150 x 40	165.1 x 48.3	G	EPDM	20
	150 x 50	165.1 x 60.3	G	EPDM	20
	150 x 65	165.1 x 76.1	G	EPDM	20
	150 x 80	165.1 x 88.9	G	EPDM	20
	150 x 100	165.1 x 114.3	G	EPDM	20
	150 x 40	168.3 x 48.3	G	EPDM	20
	150 x 50	168.3 x 60.3	G	EPDM	20
	150 x 65	168.3 x 76.1	G	EPDM	20
	150 x 80	168.3 x 88.9	G	EPDM	20
	150 x 100	168.3 x 114.3	G	EPDM	20
	200 x 80	219.1 x 88.9	G	EPDM	20
	200 x 100	219.1 x 114.3	G	EPDM	20

Continued

Grinnell Corporation (continued)

Fittings (GRUVLOK)

Type	Pipe outside diameter (mm)	Maximum working pressure (bar)
90° Elbow - Fig 7050	33.7, 42.4, 48.3, 60.3, 76.1, 88.9, 108[1], 114.3, 133[1], 159[1], 165.1, 168.3, 219.1	20
45° Elbow - Fig 7051	33.7, 42.4, 48.3, 60.3, 76.1, 88.9, 108[1],114.3, 133[1], 159[1], 165.1, 168.3, 219.1	20
Tee - Fig 7060	33.7, 42.4, 48.3, 60.3, 76.1, 88.9, 108[1] 114.3, 133[1], 159[1], 165.1, 168.3, 219.1	20
Blank Cap - Fig 7074	33.7, 42.4, 48.3, 60.3, 76.1, 88.9, 108[1] 114.3, 133[1], 159[1], 165.1, 168.3, 219.1	20

Pipe Fittings (Gruvlok)

Type	Nominal size (mm)	Pipe outside diameter (mm)	Pipe ends	Gasket type	Maximum working pressure (bar)
Style 7012	50	60.3	R&C	EPDM	20
Flange Adapter	65	76.1	R&C	EPDM	20
	80	88.9	R&C	EPDM	20
	100	114.3	R&C	EPDM	20
	125	139.7	R&C	EPDM	20
	150	165.1	R&C	EPDM	20
	150	168.3	R&C	EPDM	20
	200	219.1	R&C	EPDM	20

[1] These sizes are not suitable for installations that require BS 1387 pipework due to the sizes not being included in that standard.
[2] Two Figure 7045 mechanical tees may be used in combination with pipe run sizes 76.1mm through 168.3mm with branch sizes 48.3mm and 60.3mm, and are designated Figure 7047.
[3] Two Figure 7046 mechanical tees may be used in combination with pipe run sizes 76.1mm through 168.3mm with branch sizes 48.3mm, through 114.3mm, and are designated Figure 7048.
[4] Figure 7045 and Figure 7046 mechanical tees may be used in combination with pipe run sizes 76.1mm through 168.3mm with branch sizes 48.3mm and 60.3mm, and are designated Figure 7049.

Grinnell Sales & Distribution (Benelux)
Kopersteden 1, 7547 TJ Enschede, The Netherlands

Tel: +31 53 4283434 • Fax: +31 53 4283377

Hydrant Tee

Type	Nominal size (mm)	Pipe outside diameter (mm)	Pipe ends	Maximum working pressure (bar)
Style 7064B	100	114.3 x 76.1	76.1 BSP	20

5

SECTION 8:
8.1 PIPE COUPLINGS AND FITTINGS

Minimax GmbH

Industriestrasse 10/12, D-23840 Bad Oldesloe, Germany

Tel: +49 4531 803 0 • Fax: +49 4531 803 248
UK Sales Enquiries: Tel: +44 (0)181 832 2000 • Fax: +44 (0)181 832 2200

Sprinkler Tees

Type	Nominal size (mm)	Pipe outside diameter (mm)	Gasket type	Maximum working pressure (bar)
TXG	32 x 10	42.4	EPDM	16
TXG	32 x 15	42.4	EPDM	16
TXG	32 x 20	42.4	EPDM	16
TXG	40 x 15	48.3	EPDM	16
TXG	40 x 20	48.3	EPDM	16
TXG	50 x 20	60.3	EPDM	16
TXG	25 x 15	33.7	EPDM	16
TXG	25 x 20	33.7	EPDM	16
TXG	25 x 25	33.7	EPDM	16
TXG	32 x 25	42.2	EPDM	16
TXG	40 x 25	48.3	EPDM	16
TXG	50 x 15	60.3	EPDM	16
TXG	50 X 25	60.3	EPDM	16
TXG	65 X 20	76.1	EPDM	16

Modgal Limited

Rosh Pina Industrial Park, Upper Galilee, Israel

Tel: +972 6 691 4222 • Fax: +972 6 691 4202

Pipe Couplings (QUICKCOUP)

Type	Nominal size (mm)	Pipe outside diameter (mm)	Pipe ends	Gasket type	Maximum working pressure (bar)
Style 75	32	42.2	R & C	EPDM	20
	40	48.3	R & C	EPDM	20
	50	60.3	R & C	EPDM	20
	65	76.1	R & C	EPDM	20
	80	88.9	R & C	EPDM	20
	100	114.3	R & C	EPDM	20
	125	139.7	R & C	EPDM	20
	125	141.3	R & C	EPDM	20
	150	165.1	R & C	EPDM	20
	150	168.3	R & C	EPDM	20
	200	219.1	R & C	EPDM	20
Style 07	50	60.3	R & C	EPDM	20
	80	88.9	R & C	EPDM	20
	100	114.3	R & C	EPDM	20
	125	141.3	R & C	EPDM	20
	150	168.3	R & C	EPDM	20
	200	219.1	R & C	EPDM	20
Style 007	32	42.4	R & C	EPDM	12
Quikhinge	40	48.3	R & C	EPDM	12
	50	60.3	R & C	EPDM	12
	65	76.1	R & C	EPDM	12
	80	88.9	R & C	EPDM	12
	100	114.3	R & C	EPDM	12
	125	139.7	R & C	EPDM	12
	150	165.1	R & C	EPDM	12
	150	168.3	R & C	EPDM	12

Continued

Modgal Limited (continued)

Type	Nominal size (mm)	Pipe outside diameter (mm)	Pipe ends	Gasket type	Maximum working pressure (bar)5
Style 08 Quik-T	50	60.3	T(½ in BSP)	EPDM	20
	50	60.3	T(¾ in BSP)	EPDM	20
	50	60.3	T(1 in BSP)	EPDM	20
	50	60.3	T(1¼ in BSP)	EPDM	20
	50	60.3	T(1½ in BSP)	EPDM	20
	65	76.1	T(½ in BSP)	EPDM	20
	65	76.1	T(¾ in BSP)	EPDM	20
	65	76.1	T(1 in BSP)	EPDM	20
	65	76.1	T(1¼ in BSP)	EPDM	20
	65	76.1	T(1½ in BSP)	EPDM	20
	80	88.9	T(½ in BSP)	EPDM	20
	80	88.9	T(¾ in BSP)	EPDM	20
	80	88.9	T(1 in BSP)	EPDM	20
	80	88.9	T(1¼ in BSP)	EPDM	20
	80	88.9	T(1½ in BSP)	EPDM	20
	80	88.9	T(2 in BSP)	EPDM	20
	100	114.3	T(½ in BSP)	EPDM	20
	100	114.3	T(¾ in BSP)	EPDM	20
	100	114.3	T(1 in BSP)	EPDM	20
	100	114.3	T(1¼ in BSP)	EPDM	20
	100	114.3	T(1½ in BSP)	EPDM	20
	100	114.3	T(2 in BSP)	EPDM	20
	150	165.1	T(1¼ in BSP)	EPDM	20
	150	165.1	T(1½ in BSP)	EPDM	20
	150	165.1	T(2 in BSP)	EPDM	20
	150	168.3	T(1¼ in BSP)	EPDM	20
	150	168.3	T(1½ in BSP)	EPDM	20
	150	168.3	T(2 in BSP)	EPDM	20
	150	168.3	T(4 in BSP)	EPDM	20
Style 08 Quick-T	125 x 32	139.7 x 42.4	T	EPDM	20
	150 x 65	165.1 x 76.1	T	EPDM	20
Style 08 Quick-T	50 x 40	60.3 x 48.3	G	EPDM	20
	65 x 40	76.1 x 48.3	G	EPDM	20
	80 x 32	88.9 x 42.4	G	EPDM	20
	80 x 40	88.9 x 48.3	G	EPDM	20
	80 x 50	88.9 x 60.3	G	EPDM	20
	100 x 32	114.3 x 42.4	G	EPDM	20
	100 x 40	114.3 x 48.3	G	EPDM	20
	100 x 50	114.3 x 60.3	G	EPDM	20
	100 x 65	114.3 x 76.1	G	EPDM	20
	125 x 65	139.7 x 76.1	G	EPDM	20
	150 x 40	165.1 x 48.3	G	EPDM	20
	150 x 50	165.1 x 60.3	G	EPDM	20
	150 x 40	168.3 x 48.3	G	EPDM	20
	150 x 50	168.3 x 60.3	G	EPDM	20
	150 x 65	168.3 x 76.1	G	EPDM	20
	150 x 100	168.3 x 114.3	G	EPDM	20
Style 007R Quikhinge	32	42.2	R & C	EPDM	12
	40	48.3	R & C	EPDM	12
	50	60.3	R & C	EPDM	12
	65	76.1	R & C	EPDM	12
	80	88.9	R & C	EPDM	12
	100	114.3	R & C	EPDM	12
	125	139.7	R & C	EPDM	12
	150	165.1	R & C	EPDM	12
	150	168.3	R & C	EPDM	12

Fittings

Type	Pipe outside diameter (mm)	Maximum working pressure (bar)
Style 06 90° Elbow	42.2, 48.3, 60.3, 76.1, 88.9, 114.3, 141.3, 165.1, 168.3, 219.1	20
Style 04 45° Elbow	60.3, 76.1, 88.9, 114.3, 139.7, 165.1, 168.3, 219.1	20
Style 05 Grooved Tee	48.3, 60.3, 76.1, 88.9, 114.3, 139.7, 165.1, 168.3, 219.1	20
Style 06 90˙ Elbow	139.7	12
Style 02/020 Blank Ends	42.2, 48.3, 60.3, 76.1, 88.9, 114.3, 139.7, 165.1, 168.3, 219.1	20
Style 02 Blank End	42.4, 48.3, 76.1, 88.9, 139.7, 168.3	20

163

Victaulic Company of America

4901 Kesslersville Road, Easton, Pennsylvania 18040, USA

Tel: +1 610 559 3300 • Fax: +1 610 250 8817
Sales Enquiries: Victaulic Europe
Tel: +32 9 369 4454 • Fax: +32 9 369 7518, 9 366 2553

Pipe Couplings

Type	Nominal size (mm)	Pipe outside diameter (mm)	Pipe ends (mm)	Gasket type	Maximum working pressure (bar)
Style 005	50	60.3	R & C	EPDM	12
		73.0[2]	R & C	EPDM	12
	65	76.1	R & C	EPDM	12
	80	88.9	R & C	EPDM	12
		108.0[2]	R & C	EPDM	12
	100	114.3	R & C	EPDM	12
		133.4[2]	R & C	EPDM	12
	125	139.7	R & C	EPDM	12
		141.3[2]	R & C	EPDM	12
		158.8[2]	R & C	EPDM	12
	150	165.1	R & C	EPDM	12
	150	168.3	R & C	EPDM	12
	200	219.1	R & C	EPDM	12
Style 07	25	33.4	R & C	EPDM	20
	32	42.2	R & C	EPDM	20
	40	48.3	R & C	EPDM	20
	50	60.3	R & C	EPDM	20
		73.0[2]	R & C	EPDM	20
	65	76.1	R & C	EPDM	20
	80	88.9	R & C	EPDM	20
		108.0[2]	R & C	EPDM	20
	100	114.3	R & C	EPDM	20
		133.4[2]	R & C	EPDM	20
	125	139.7	R & C	EPDM	20
		141.3[2]	R & C	EPDM	20
		158.8[2]	R & C	EPDM	20
	150	165.1	R & C	EPDM	20
	150	168.3	R & C	EPDM	20
	200	219.1	R & C	EPDM	20
Style 75	40	48.3	R & C	EPDM	20
	50	60.3	R & C	EPDM	20
	80	88.9	R & C	EPDM	20
	100	114.3	R & C	EPDM	20
	125	139.7	R & C	EPDM	20
	150	165.1	R & C	EPDM	20
	200	219.1	R & C	EPDM	20
Style 77	20	26.7	R & C	EPDM	20
	25	33.4	R & C	EPDM	20
	32	42.2	R & C	EPDM	20
	40	48.3	R & C	EPDM	20
	50	60.3	R & C	EPDM	20
	65	76.2	R & C	EPDM	20
	80	88.9	R & C	EPDM	20

[1] The above range of couplings (i.e. style 005 and 07) are branded 'VICTAULIC', 'PJE', or 'PIPECO'.
[2] These sizes are not suitable for installations that require BS 1387 pipework due to the sizes not being included in that standard.

Continued

Victaulic Company of America (continued)

Type	Nominal size (mm)	Pipe outside diameter (mm)	Pipe ends (mm)	Gasket type	Maximum working pressure (bar)
Style 77	100	114.3	R & C	EPDM	20
	150	165.1	R & C	EPDM	20
	200	219.1	R & C	EPDM	20
Style 750	50 x 40	60.3 x 48.3	R & C	EPDM	20
Reducing	80 x 50	88.9 x 60.3	R & C	EPDM	20
	100 x 50	114.3 x 60.3	R & C	EPDM	20
	100 x 80	114.3 x 88.9	R & C	EPDM	20
	150 x 100	165.1 x 114.3	R & C	EPDM	20
	200 x 150	219.1 x 165.1	R & C	EPDM	20
Mechanical Tees					
Victaulic	100 x 50	114.3 x 60.3	G	EPDM	20
Mechanical	100 x 80	114.3 x 88.9	G	EPDM	20
Tees	150 x 50	165.1 x 60.3	G	EPDM	15
	150 x 80	165.1 x 88.9	G	EPDM	15
	150 x 100	165.1 x 114.3	G	EPDM	15
	200 x 80	219.1 x 88.9	G	EPDM	20
	200 x 100	219.1x 114.3	G	EPDM	20
Style 927	32 x 15	42.4 x 21.3	T	EPDM	12
	32 x 20	42.4 x26.7	T	EPDM	12
	32 x 25	42.4 x 26.7	T	EPDM	12
	40 x 15	48.3 x21.3	T	EPDM	12
	40 x 20	48.3 x 26.7	T	EPDM	12
	40 x 25	48.3 x 33.7	T	EPDM	12
	50 x 15	60.3 x 21.3	T	EPDM	12
	50 x 20	60.3 x 26.7	T	EPDM	12
	50 x 25	60.3 x 33.7	T	EPDM	12
Style 741 (Vic adaptors)					
"	80	88.9	R & C	EPDM	20
"	125	139.7	R & C	EPDM	20
"	150	165.1	R & C	EPDM	20
"	150	168.3	R & C	EPDM	20
"	200	219.1	R & C	EPDM	20
Style 920	80x32	89x42	T	EPDM	20
	80x40	89x48	T	EPDM	20
	100x32	114x42	T	EPDM	20
	100x40	114x48	T	EPDM	20
	150x40	165x48	T	EPDM	20
	150x80	165x89	T	EPDM	20
	150x80	168x89	T	EPDM	20
	200x80	219x89	T	EPDM	20
	80x40	89x48	G	EPDM	20
	100x40	114x48	G	EPDM	20
	100x80	114x89	T	EPDM	20
	150x100	165x114	T	EPDM	20
	150x100	168x114	T	EPDM	20
	200x65	219x76	T	EPDM	20
	200x100	219x114	T	EPDM	20
Style 75	65	76.1	R & C	EPDM	20

Continued

SECTION 8:

8.1 PIPE COUPLINGS AND FITTINGS

Victaulic Company of America (continued)

Fittings

Type	Nominal pipe size (mm)	Maximum working pressure (bar)
90° Elbow	20, 25, 32, 40, 50, 65, 80, 100, 150, 200	20
90° Long Radius Elbow	50, 80, 100, 150, 200	20
45° Elbow	20, 25, 32, 40, 50, 80, 100, 150, 200	20
45° Long Radius Elbow	50, 80, 100, 150, 200	20
22.5° Elbow	25, 32, 40, 50, 80, 100, 150, 200	20
11.25° Elbow	25, 32, 40, 50, 80, 100, 150, 200	20
Tee	20, 25, 32, 40, 50, 65, 80, 100,150, 200	20
Cross	20, 25, 32, 40, 50, 80, 100, 150, 200	20
Cap	20, 25, 32, 40, 50, 80, 100, 150, 200	20
45° Lateral	20, 25, 32, 40, 50, 80, 100, 150, 200	20
90° True Y	25, 32, 40, 50, 80, 100, 200	20
Reducer	40 x 25, 50 x 20, 50 x 25, 50 x 32, 50 x 40, 80 x 25, 80 x 50, 100 x 125, 100 x 40, 100 x 50, 100 x 80, 200 x 80, 200 x 100	20
45° Reducing Lateral	80 x 80 x 50, 100 x 100 x 50, 100 x 100 x 80, 200 x 200 x 100	20
Reducing Tee	50 x 50 x 20, 50 x 50 x 25, 50 x 50 x 40, 80 x 80 x 25, 80 x 80 x 40, 80 x 80 x 50, 100 x 100 x 25, 100 x 100 x 50, 100 x 100 x 80, 150 x 150 x 50, 150 x 150 x 100, 200 x 200 x 50, 200 x 200 x 80, 200 x 200 x 100, 200 x 200 x 150	20

All couplings and fittings to be installed in accordance with the Victaulic Field Assembly and Installation Instruction handbook.

Viking Johnson (incorporating Victaulic Systems)
PO Box 13, 46/48 Wilbury Way, Hitchin, Herts SG4 0UD

Tel: +44 (0)1462 422622 • Fax: +44 (0)1462 422072

Pipe Couplings

Type	Nominal size (mm)	Pipe outside diameter (mm)	Pipe ends	Gasket type	Maximum working pressure (bar)
Style 75 Standard joint					
	20	26.9	C	EPDM	20
	40	48.4	C	EPDM	20
	50	60.3	R & C	EPDM	20
	65	76.1	R	EPDM	20
	80	88.9	R	EPDM	20
	100	114.3	R	EPDM	20
	125	139.7	R	EPDM	15
	125	141.3	R	EPDM	15
	150	165.1	R	EPDM	15
	150	168.3	R	EPDM	15
	200	219.1	R & C	EPDM	20
Style 77 Heavy duty					
Pattern No. 6527M	25	33.7	R & C	EPDM	20
Pattern No. 6528M	32	42.4	R & C	EPDM	20
Pattern No. 6492M	40	48.3	R & C	EPDM	20
Style 750	65 x 50	76.1 x 60.3	R & C	EPDM	20
Reducing	80 x 65	88.9 x 76.1	R & C	EPDM	20
	100 x 65	114.3 x 76.1	R & C	EDPM	16
	100 x 80	114.3 x 88.9	R & C	EPDM	20
	150 x 80	165.1 x 88.9	R & C	EPDM	15
	150 x 100	219.1 x 165.1	R & C	EPDM	15
Solid Vic-Adaptors					
Style 5510	50	60.3	R	EPDM	16
	80	88.9	R	EPDM	16
	100	114.3	R	EPDM	16
	150	165.1	R	EPDM	16
	150	168.3	R	EPDM	16

Continued

Viking Johnson (continued)

Type	Nominal size (mm)	Pipe outside diameter (mm)	Pipe ends	Gasket type	Maximum working pressure (bar)
Hinged Vic-Adaptor					
Style 5515	80	88.9	R & C	EPDM	16
	125	139.7	R & C	EPDM	16
	150	165.1	R & C	EPDM	16
	150	168.3	R & C	EPDM	16
	200	219.1	R & C	EPDM	16
Mechanical Tees					
Style 920	65	76.1	T (1.5 in BSP)	EPDM	20
	80	88.9	T (2 in BSP)	EPDM	20
	100	114.3	T (2 in BSP)	EPDM	20
	100	114.3	T (2.5 in BSP)	EPDM	20
	150/150	168.3/165.1	T (2 in BSP)	EPDM	20
	100 x 50	114.3 x 60.3	G	EPDM	20
	100 x 65	114.3 x 76.1	G	EPDM	20
	100 x 80	114.3 x 88.9	G	EPDM	20
	150/150 x 50	168.3/165.1 x 60.3	G	EPDM	20
	150/150 x 65	168.3/165.1 x 76.1	G	EPDM	20
	150/150 x 80	168.3/165.1 x 88.9	G	EPDM	20
	150/150 x 100	168.3/165.1 x 114.3	G	EPDM	20
	200 x80	219.1 x 88.9	G	EPDM	20
	200 x 100	219.1 x 114.3	G	EPDM	20

Fittings

Type	Nominal pipe size (mm)	Maximum working pressure (bar)
90° Elbow	40, 50, 65, 80, 100, 125, 150, 200	20
45° Elbow	40, 50, 65, 80, 100, 125, 150, 200	20
Tee	40, 50, 65, 80, 100, 125, 150, 200	20
Blank End	25, 32, 40, 50, 65, 80, 100, 125, 150, 200	20
Grooved (Radial Branch)		
Pitcher Tee	100 x 100 x 2.5 in BSP	20

All couplings and fittings to be installed in accordance with 'The Victaulic Grooved Pipe Jointing System' Catalogue.

Virotec Rohrtechnik GmbH & Co KG
Rudolf Diesel Strasse 4, 63571 Gelnhausen, Germany
Tel: +49 6051 4819-0 • Fax: +49 6051 3850

Type	Nominal size (mm)	Pipe outside diameter (mm)	Gasket type	Maximum working pressure (bar)
Model V60	25 x 10	33.7	Silicone	12
"	25 x 15	33.7	Silicone	12
"	25 x 20	33.7	Silicone	12
"	25 x 25	33.7	Silicone	12
Model V100	25 x 10	33.7	Silicone	12
"	25 x 15	33.7	Silicone	12
"	25 x 20	33.7	Silicone	12
"	25 x 25	33.7	Silicone	12
Sprinkler Tees	25 x 10	33.7	EPDM	12
"	25 x 15	33.7	EPDM	12
"	25 x 20	33.7	EPDM	12
"	25 x 25	33.7	EPDM	12
"	32 x 10	42.2	EPDM	12
"	32 x 15	42.2	EPDM	12
"	32 x 20	42.2	EPDM	12
"	32 x 25	42.2	EPDM	12
"	40 x 10	48.3	EPDM	12
"	40 x 15	48.3	EPDM	12

Continued

Virotec Rohrtechnik GmbH (continued)

Type	Nominal size (mm)	Pipe outside diameter (mm)	Gasket type	Maximum working pressure (bar)
Sprinkler Tees				
"	40 x 20	48.3	EPDM	12
"	40 x 25	48.3	EPDM	12
"	50 x 10	60.3	EPDM	12
"	50 x 15	60.3	EPDM	12
"	50 x 20	60.3	EPDM	12
"	50 x 25	60.3	EPDM	12
Fittings				
Adjustable multiple angle pipe fittings				
Flexarm	25 x 25	33.7	Silicone	12
Triflexarm	25 x 10	33.7	Silicone	12
	25 x 15	33.7	Silicone	12
	25 x 20	33.7	Silicone	12

Model	Adjustment (without shortening)		Nominal Size /Connection		Max. Working Pressure (bar)
	Min Length (mm)	Max. Length (mm)	Supply	Outlet	
VL 0.5	400	520	R1"ext.thread	Rp3/8" int.thread	12
			R1"ext.thread	Rp1/2" int.thread	
			R1"ext.thread	Rp3/4" int.thread	
VL 1.0	880	1000	R1"ext.thread	Rp3/8" int.thread	12
			R1"ext.thread	Rp1/2" int.thread	
			R1"ext.thread	Rp3/4" int.thread	
VLS 0.5	400	500	R1"ext.thread	Rp3/8" int.thread	12
			R1"ext.thread	Rp1/2" int.thread	
VLS 1.0	880	980	R1"ext.thread	Rp3/8" int.thread	12
			R1"ext.thread	Rp1/2" int.thread	

1. The length of the pipe is adjustable by pushing/pulling (VL versions) or screwing (VLS versions). The length can also be changed by shortening (cutting off) of the parts to a minimum length of 220mm, as specified by the manufacturer.
2. The product is further described in Virotec document 'Telescope Pipe Models VL and VLS - Product Description'.

Witzenmann GmbH*

Östl. Karl-Friedrich-Strasse 134, D-75175 Pforzheim, Postfach 1280, D-75112 Pforzheim, Germany

Tel: +49 7231 581 0 • Fax: +49 7231 581 820

Hydra stainless steel flexible pressure hoses

Hydra RS 331 L 12/RS 321 L 12 threaded connections DN16, DN20, DN25, DN32, DN40, DN50[1]
Hydra RS 331 L 12/RS 32I L 12 flanged connections DN20, DN25, DN32, DN40, DN 50[1]
Hydra RS 330 L 12 flanged connections DN65, DN80, DN100[1]

1. The maximum working pressure for the ranges is 20 bar.

* This company has not been certified by the LPCB to ISO 9000.

SECTION 9: PRE-ACTION SYSTEMS
9.1 CONTROL EQUIPMENT

Angus Fire Armour Limited
Kastanievej 15, DK-5620, Glamsbjerg, Denmark

UK Sales Enquiries: Tel: +44 (0)184 421 4545 • Fax: +44 (0)184 421 3511

ANGUS PA MK III TYPE 1 PRE-ACTION SYSTEM
Approved Pre-Action Control Valve Description

Size(s)	Mounting Attitude	Body Material	Model	Type	Maximum Working Pressure	Flanged or Grooved
100mm	Vertical	Cast iron	MKIII	Pre-Action 100	12 bar	Flanged
150mm	Vertical	Cast iron	MKIII	Pre-Action 150	12 bar	Flanged

Approved Solenoid Valve Description

Manufacturer	Model	Voltage (V)	Power (W)
Danfoss	EVSI 15 (nc)	24 DC	15

Approval Conditions

1 The system employs duplicate solenoid valves, which are manufactured by Danfoss, as detailed above.
2 Solenoid valves shall be installed in parallel.
3 Solenoid valves shall function in pneumatic condition only, and be protected by a strainer.
4 LPCB Certificated detectors compatible with the control and indicating equipment shall be used.
5 Suitable electrical detection control and indicating equipment and pneumatic systems which are compatible with the equipment above shall be used.
   The control and indicating equipment shall be LPCB approved/certificated.
6 Connecting cables shall comply with BS 6387 : 1994, classification C, W, Z.
7 The system shall be electrically monitored to demonstrate that it is in a ready-to-operate state at all times.
8 Clean dry air shall be used.
9 The equipment must be installed, operated and maintained as prescribed in Angus Fire Ltd. installation, operating and maintenance manual.
10 The pre-action system shall be configured in accordance with Angus specification numbers, PA1/100 and PA1/150 for the 100mm and 150mm versions respectively.
11 The Pre-Action systems shall comply with the details specified in LPC Technical Bulletin 21 *Supplementary requirements for sprinkler installations which can operate in the dry mode.*
12 Angus Fire, Denmark, is responsible for the complete pre-action system.

Grinnell Manufacturing (UK) Limited
Stockport Trading Estate, Yew Street, Stockport, SK4 2JW

Tel: +44 (0)161 477 1886 • Fax: +44 (0)161 477 6729

Mark I pre-action control equipment (Types 1 and 2)
Grinnell Mark III TYPE 1 Pre-action system

Approved Pre-action Control Valve Description

Size(s)	Mounting Attitude	Body Material	Model	Type	Maximum Working Pressure	Flanged or Grooved
100mm	Vertical	Gunmetal	FE 2349	Alternate	16 bar	Wafer type
150mm	Vertical	Gunmetal	FE 2152	Alternate	16 bar	Wafer type
200mm	Vertical	Gunmetal	FE 2380	Alternate	16 bar	Wafer type

Approved Solenoid Valve Description

Manufacturer	Model	Voltage (V)	Power (W)
Skinner	LV2LBX25	24 DC	11
ASCO	R8210A107	24 DC	16.8

Continued

Grinnell Manufacturing (UK) Limited (continued)

Approval Conditions

1   The above pre-action system employs duplicate solenoid valves, which are manufactured by Skinner, and ASCO, as detailed above. One of each shall be used.
2   Solenoid valves shall be installed in parallel.
3   Solenoid valves shall function in pneumatic conditions only, and be protected by a strainer.
4   LPCB Certificated detectors compatible with the control and indicating equipment shall be used.
5   Suitable electrical detection control and indicating equipment and pneumatic systems which are compatible with the equipment identified above shall be used. The control and indicating equipment shall be LPCB approved/certificated.
6   Connecting cables shall comply with BS 6387: 1994, classification C, W, Z evidenced by LPCB certification.
7   The system shall be electrically monitored to demonstrate that it is in a 'ready-to-operate' state at all times.
8   Clean dry air shall be used.
9   The equipment must be installed, operated and maintained as prescribed in Wormald Manufacturing Ltd. installation, operating and maintenance manual.
10   The pre-action system shall be configured in accordance with Wormald specification number 03598, 03599 and 03600 for the 100mm, 150mm and 200mm versions respectively.
11   The pre-action systems shall comply with the details specified in LPC Technical Bulletin 21 *Supplementary Requirements for sprinkler installations which can operate in dry mode.*
12   Wormald Manufacturing Ltd. are the firm responsible for the complete pre-action station, including electrics.

The Reliable Automatic Sprinkler Co Inc
525 North MacQuesten Parkway, Mount Vernon, New York 10552-2600, USA

Tel: +1 914 668 3470 • Fax: +1 914 668 2936
UK Sales Enquiries: Tel: +44 (0)1372 728899 • Fax: +44 (0)1372 724461

RELIABLE SUPERTROL TYPE 1 PRE-ACTION SYSTEM

a. Approved Pre-action Control Valve Description

Size(s)	Mounting Attitude	Body Material	Model	Type	Maximum Working Pressure	Flanged or Grooved
100mm	Vertical	Cast Iron	B/E	Pre-action	12 bar	Flanged
150mm	Vertical	Cast Iron	B/E	Pre-action	12 bar	Flanged

b. Approved Solenoid Valve Description

Manufacturer	Model	Voltage (V)	Power (W)
Skinner	LV2LBX25	24DC	11
ASCO	R8210A107	24DC	16.8

Approval Conditions

1.   The above pre-action system employs duplicate solenoid valves, which are manufactured by firms, as detailed above. One of each shall be used.
2.   Solenoid valves shall be installed in parallel.
3.   Solenoid valves shall function in pneumatic condition only, and be protected by a strainer.
4.   LPCB Certificated detectors compatible with the control and indicating equipment shall be used.
5.   Suitable electrical detection control and indicating equipment and pneumatic systems which are compatible with the equipment identified above shall be used. The control and indicating equipment should be LPCB approved/certificated.
6.   Connecting cables shall comply with BS 6387: 1994, classification C, W, Z evidenced by LPCB certification.
7.   The system shall be electrically monitored to demonstrate that it is in a 'ready-to-operate' state at all times.
8.   Clean dry air shall be used.
9.   The equipment must be installed, operated and maintained as prescribed in The Reliable Automatic Sprinkler Co Inc installation, operating and maintenance manual.
10.   The pre-action system shall be configured in accordance with Reliable specifications.
11.   The pre-action systems shall comply with the details specified in LPC Technical Bulletin 21 Supplementary Requirements for sprinkler installations which can operate in the dry mode and in the LPC Rules for Automatic Sprinkler Installations.
12.   The Reliable Automatic Sprinkler Company Inc are the firm responsible for the complete pre-action station, including electrics. It should be noted that only the equipment specifically referenced above under (a) and (b) is LPCB approved.

The Viking Corporation
210 N Industrial Park Road, Hastings, Michigan 49058, USA

Tel: +1 616 945 9501 • Fax: +1 616 945 9599
UK Sales Enquiries: Tel: +44 (0)1427 875999 • Fax: +44 (0)1427 875998

Viking Firecycle system incorporating the following equipment:
    Flow control valves (Model H-1) 50mm 80mm 100mm 150mm
    Control panel (Model B-1)
    Heat detectors (Type FC 1002 - 140'F)
    M1 cable used for connecting the heat detectors in a series circuit

SECTION 10: 5
*P*UMP STARTING AND ELECTRICAL ALARM PRESSURE SWITCHES

The following pressure switches have been approved for use under the LPC *Rules for automatic sprinkler installations.*

Bailey & Mackey Limited
Baltimore Road, Birmingham B42 1DE

Tel: +44 (0)121 357 5351 • Fax: +44 (0)121 357 8319

LPCB Ref. No.

085a/01	Bailey & Mackey	Pressure Switch Type 1381, 1381V, 1381F
085a/02	Bailey & Mackey	Pressure Switch Type 108, 108V

Potter Electric Signal Co.
2081 Craig Road, St Louis, Missouri 63146, USA

Tel: +1 314 878 4321 • Fax: +1 314 878 7264

Model PS 10A
Model PS 40A
Model PS 100A
Model PS 120A

INTRODUCTION

The *LPC Rules for automatic sprinkler installations,* now includes Technical Bulletin TB 20: *The Selection of Sprinkler Heads,* which shall be used to select sprinkler heads according to their thermal sensitivity and other characteristics specified in the Technical Bulletin, as appropriate to the hazard classification and type of risk. Technical Bulletin TB20 should be used in conjunction with BS 5306: Part 2: 1990, clause 25, *Sprinkler, multiple control and sprayer design characteristics and uses.*

The Thermal Sensitivity Ratings in TB 20 are: Quick, Special, Standard A and Unrated.

Standard B Thermal Sensitivity Rating sprinklers are obsolete.

Quick and Special Response sprinklers are identified in the listings. Classification of Standard A and deletion of Standard B sprinklers is now planned.

Recessed, concealed and sidewall (horizontal) sprinklers have an Unrated Thermal Sensitivity Rating and are suitable for risks, or part risks, as detailed in LPCB Technical Bulletin TB 20: *The Selection of Sprinkler Heads.*

KEY to abbreviations

Each entry includes the nominal orifice size in millimetres and the temperature ratings in degrees centigrade.

† indicates glass bulb sprinkler head; all others are fusible element sprinkler heads.

* indicates that a corrosion-resistant coating for the body and yoke of the sprinkler has been approved under the LPC *Rules for Automatic Sprinkler Installations.*
The type of sprinkler is indicated by the following abbreviations:

Conventional pattern sprinkler (upright & pendent)	CU/P	Conventional pattern dry sprinkler (upright & pendent)	CDU/P
Conventional pattern sprinkler (upright)	CU	Conventional pattern dry sprinkler (upright)	CDU
Conventional pattern sprinkler (pendent)	SP	Conventional pattern dry sprinkler (pendent)	CDP
Spray pattern sprinkler (upright)	SU	Spray pattern dry sprinkler (upright)	SDU
Spray pattern sprinkler (pendent)	SP	Spray pattern dry sprinkler (pendent)	SDP
Sidewall sprinkler (upright & pendent)	WU/P	Conventional pattern ceiling flush sprinkler	CF
Sidewall sprinkler (upright)	WU	Spray pattern ceiling flush sprinkler	SF
Sidewall sprinkler (pendent)	WP	Spray pattern ceiling flush dry sprinkler	SFD
Sidewall sprinkler (horizontal)	WH	Spray pattern recessed sprinkler	SR
Sidewall sprinkler flush (horizontal)	WHF	Spray pattern concealed sprinkler	SK
Sidewall sprinkler extended coverage (horizontal)	WHEC	Early suppression fast response sprinkler	ESFR

AFAC Inc.

PO Box 400, West 2nd Avenue, Ranson, West Virginia 25438, USA

Tel: +1 304 728 9000 • Fax: +1 304 728 8481

† 'Automatic' Model H[1,2]	15mm	CU/P	57° 68° 79° 93° 141° 182°
† 'Automatic' Model H[1,2]	15mm	SU	57° 68° 79° 93° 141° 182°
† 'Automatic' Model H[1,2]	15mm	SP	57° 68° 79° 93° 141° 182°
† 'Automatic' Model H[1,2]	20mm	SU	57° 68° 79° 93° 141° 182°
† Model K1 ESFR	20mm	SP	68°, 93°
† Model HC[3,4]	15mm	SK	(adjustable) 68° with 57° ceiling plate (5mm bulb)
† Model HC[3,4]	15mm	SK	(adjustable) 68° with 57° ceiling plate (3mm bulb)
† Model HC[3,4]	15mm	SK	(adjustable) 93° with 74° ceiling plate (5mm bulb)
† Model HC[3,4]	15mm	SK	(adjustable) 93° with 74° ceiling plate (3mm bulb)

[1] Also approved with a chromium finish.
[2] The 68°, 79° and 93° sprinklers are also approved with wax and wax over lead finishes.
[3] Approved with chrome and coloured coverplate.
[4] To be installed in accordance with specific installation instructions. Not to be installed in ceilings
   with a positive pressure in the void above. The ceiling plate is not to be repainted.

Angus Fire Armour Limited

Kastanievej 15, DK-5620, Glamsbjerg, Denmark

UK Sales Enquiries: Tel: +44 (0)1844 214545 • Fax: +44 (0)1844 213511

† Angus R-F2[2,4,5]	15mm	WHEC	57°, 68°, 79°
† Angus S(7)[1,2]	10mm	SR	57°, 68°, 79°, 93°, (5mm bulb)
† Angus S(7)[2]	10mm	SP	57°, 68°, 79°, 93°, 141°, 182°, (5mm bulb)
† Angus S(8)[2]	10mm	SU	57°, 68°, 79°, 93°, 141°, 182°, (5mm bulb)
† Angus S(4)[1,2,7]	10mm	SR	57°, 68°, 79°, 93°, (5mm bulb)
† Angus S(4)[2,7]	10mm	SP	57°, 68°, 79°, 93°, 141°, 182°, (5mm bulb)
† Angus S(6)[2,7]	10mm	SU	57°, 68°, 79°, 93°, 141°, 182°, (5mm bulb)
† Angus S(1)[2]	15mm	CU/P	57°, 68°, 79°, 93°, 141°, 182°, (5mm bulb)
† Angus S(2)[2]	15mm	SR	57°, 68°, 79°, 93°,(5mm bulb)
† Angus S(2)[2]	15mm	SP	57°, 68°, 79°, 93°, 141°, 182°, (5mm bulb)
† Angus S(3)[2]	15mm	SU	57°, 68°, 79°, 93°, 141°, 182°, (5mm bulb)
† Angus S(5)[2]	15mm	WU/P	57°, 68°, 79°, 95°, 141°, 182° (5mm bulb)
† Angus S(14)[2,6]	20mm	CU/P	57°, 68°, 79°, 93°, 141°, 182°, (5mm bulb)
† Angus S(12)[2,6]	20mm	SP	57°, 68°, 79°, 93°, 141°, 182°, (5mm bulb)
† Angus S(13)[2,6]	20mm	SU	57°, 68°, 79°, 93°, 141°, 182°, (5mm bulb)
† Angus S(9)[2]	20mm	SP	57°, 68°, 79°, 93°, 141°, 182°, (5mm bulb)
† Angus S(10)[2]	20mm	SU	57°, 68°, 79°, 93°, 141°, 182°, (5mm bulb)
† Angus S(11)[2]	20mm	CU/P	57°, 68°, 79°, 93°, 141°, 182°, (5mm bulb)
† Angus S(7)QR[1,2]	10mm	SR	57°, 68°, 79°, 93°, 141°
† Angus S(7)QR[2,3]	10mm	SP	57°, 68°, 79°, 93°, 141° (quick response)
† Angus S(8)QR[2,3]	10mm	SU	57°, 68°, 79°, 93°, 141° (quick response)
† Angus S(2)QR[1,2]	15mm	SR	57°, 68°, 79°, 93°, 141°
† Angus S(2)QR[2,3]	15mm	SP	57°, 68°, 79°, 93°, 141° (quick response)
† Angus S(3)QR[2,3]	15mm	SU	57°, 68°, 79°, 93°, 141° (quick response)
† Angus S(1)QR[2,3]	15mm	CU/P	57°, 68°, 79°, 93°, 141° (quick response)
† Angus S(5)QR[2,3]	15mm	WU/P	57°, 68°, 79°, 93°, 141° (quick response)
† Angus S(4)QR[1,2,7]	10mm	SR	57°, 68°, 79°, 93°, 141°
† Angus S(4)QR[2,3,7]	10mm	SP	57°, 68°, 79°, 93°, 141° (quick response)
† Angus S(6)QR[2,3,7]	10mm	SU	57°, 68°, 79°, 93°, 141° (quick response)
† Angus S(12)QR[2,3,6]	20mm	SP	57°, 68°, 79°, 93°, 141° (quick response)
† Angus S(14)QR[2,3,6]	20mm	CU/P	57°, 68°, 79°, 93°, 141° (quick response)
† Angus S(13)QR[2,3,6]	20mm	SU	57°, 68°, 79°, 93°, 141° (quick response)
† Angus S(9)QR[2,3]	20mm	SP	57°, 68°, 79°, 93°, 141° (quick response)
† Angus S(10)QR[2,3]	20mm	SU	57°, 68°, 79°, 93°, 141° (quick response)
† Angus S(11)QR[2,3]	20mm	CU/P	57°, 68°, 79°, 93°, 141° (quick response)
† Angus HC[8,9]	15mm	SK	(adjustable) 68° with 57° ceiling plate (5mm bulb)
† Angus HC[8,9]	15mm	SK	(adjustable) 68° with 57° ceiling plate (3mm bulb)
† Angus HC[8,9]	15mm	SK	(adjustable) 93° with 74° ceiling plate (5mm bulb)
† Angus HC[8,9]	15mm	SK	(adjustable) 93° with 74° ceiling plate (3mm bulb)

1   Approved with brass, chrome or coloured recessing cup.
2   Also approved with chromium and painted finishes.
3   The suitability of 'quick response' sprinklers for many risks is not yet proven. Use is therefore limited to where the
    purchaser, authority having jurisdiction and/or insurers consent to its use. Quick response sprinklers must be handled with
    care due to the fragile nature of the operating element, with packaging and instructions being marked accordingly.
4   This extended coverage horizontal sidewall sprinkler should be handled with care to ensure that the deflector is not
    deformed. After installation the deflector angle should be checked with the manufacturer's gauge to verify its angle of 92.5°.
5   The Product manufacturer should be consulted to determine the product fire protection capability and installation practice.
    Available test evidence and the CEA document 'Requirements and Test Methods for Hotel Room Horizontal Sidewall
    Sprinklers' should also be consulted to ensure correct usage in light hazard installations only.
6   This 20mm orifice sprinkler is fitted with a pintle to denote that it has a 15mm nominal thread size.
7   This 10mm sprinkler is fitted with a pintle to denote that it has a 15mm nominal thread size.
8   Approved with chrome and coloured coverplate.
9   To be installed in accordance with specific installation instructions. Not to be installed in ceilings with a positive pressure in
    the void above. The ceiling plate is not to be repainted.
NOTE: The numbers in brackets are not marked on the Angus S and Angus S QR Sprinklers.

Central Sprinkler Company

Corringham Road Industrial Estate, Gainsborough, Lincolnshire DN21 1QB

Tel: +44 (0)1427 615401 • Fax: +44 (0)1427 610433

Sales Enquiries - Tel: +1 215 362 0700 • Fax: +1 215 362 5385

Spraysafe A [2, 3 ,4]	15mm	SK	72° with 57° ceiling plate, 100° with 74° ceiling plate
† Spraysafe B[5]	10mm	SP	57° 68° 79° 93° 141°
† Spraysafe B[5]	10mm	SU	57° 68° 79° 93° 141°
† Spraysafe B[5]	15mm	CU/P	57° 68° 79° 93° 141°
† Spraysafe B[5]	15mm	SU	57° 68° 79° 93° 141°
† Spraysafe B[5]	15mm	SP	57° 68° 79° 93° 141°
† Spraysafe B[2, 3, 6, 7]	15mm	WU/P	57° 68° 79° 93° 141°
† Spraysafe B[5]	15mm	SR	57° 68°
† Spraysafe BL020[8]	20mm	CU/P	68° 79° 93° 141°
† Spraysafe S10[5]	20mm	CU/P	68° 79°
† Spraysafe S10[5]	10mm	SP	57°,68°,79°,93°,141° (5mm bulb)
† Spraysafe S15[5]	10mm	SU	57°,68°,79°,93°,141° (5mm bulb)
† Spraysafe S15[5]	15mm	SP	57°,68°,79°,93°,141° (5mm bulb)
† Spraysafe S15[5]	15mm	SU	57°,68°,79°,93°,141° (5mm bulb)
† Spraysafe S15[5]	15mm	CU/P	57°,68°,79°,93°,141° (5mm bulb)
† Spraysafe S15[5]	15mm	SR	57°,68°,79° (5mm bulb)
† Spraysafe S15[5]	15mm	WU/P	57°,68°,79°,93°,141° (5mm bulb)
† Spraysafe SC10[5]	10mm	SU	68° (5mm bulb)(special response)
† Spraysafe SC10[5]	10mm	SP	68° (5mm bulb)(special response)
† Spraysafe SC15[5]	15mm	CU/P	68° (5mm bulb)(special response)
† Spraysafe SC15[5]	15mm	SU	68° (5mm bulb)(special response)
† Spraysafe SC15[5]	15mm	SP	68° (5mm bulb)(special response)
† Spraysafe SC15[5]	15mm	SR	68° (5mm bulb)
† Spraysafe SC15[5]	15mm	WU/P	68° (5mm bulb)(special response)
† Spraysafe SFR10[5,9]	10mm	SU	57°,68°,79°,93°,141° (quick response)
† Spraysafe SFR10[5,9]	10mm	SP	57°,68°,79°,93°,141° (quick response)
† Spraysafe SFR15[5,9]	15mm	CU/P	57°,68°,79°,93°,141° (quick response)
† Spraysafe SFR15[5,9]	15mm	SU	57°,68°,79°,93°,141° (quick response)
† Spraysafe SFR15[5,9]	15mm	SP	57°,68°,79°,93°,141° (quick response)
† Spraysafe SFR15[5]	15mm	SR	57°,68°,79° (3mm bulb)
† Spraysafe SFR15[5,9]	15mm	WU/P	57°,68°,79°,93°,141° (quick response)

[2] To be installed with specific installation instructions.
[3] Not to be installed in ceilings with a positive pressure in the void above.
[4] Approved with chrome and coloured ceiling plates. The ceiling plates shall not be repainted.
[5] Also approved with chrome and coloured finish.
[6] To be installed in accordance with general arrangement drawing number 4-7150097.
[7] The escutcheon plate is approved in metal with plain, chrome and coloured finish and in thermosetting plastic with plain (i.e. black & white) chrome and bronze finish.
[8] This 20mm orifice sprinkler is fitted with a pintle to denote that it has a ½" nominal thread size.
[9] The suitability of 'quick response' sprinklers for many risks is not yet proven. Use is therefore limited to where the purchaser, authority having jurisdiction and/or insurers consent to its use. Quick response sprinklers must be handled with care due to the fragile nature of the operating element, with packaging and instructions being marked accordingly.

Central Sprinkler Company

451 N. Cannon Avenue, Lansdale, Pennsylvania 19446, USA

Tel: +1 215 362 0700 • Fax: +1 215 362 5385

Central A[1]	15mm	CU/P	57° 75° 100° 141°
Central A[1]	15mm	SU	57° 75° 100° 141°
Central A[1]	15mm	SP	57° 75° 100° 141°
Central A[2]	15mm	SK	72° with 57° ceiling plate, 100° with 74° ceiling plate
Central A	20mm	SU	57° 75° 100° 141°
Central A	20mm	SP	57° 75° 100° 141°
Central ESFR-1	20mm	SP	71° 100°

[1] Also approved with a chromium finish.
[2] Approved with chrome and white ceiling plates. To be installed in accordance with specific installation instructions. Not to be installed in ceilings with a positive pressure in the void. The ceiling plate is not to be repainted.

CPF Industriale SpA

Via E. Fermi, 98, 25064 Gussago, (Brescia), Italy

Tel: +39 030 310461 • Fax: +39 030 310465

† CPF SKR	15mm	CU/P	57°,68°,79°,93°,141°,182° (5mm bulb)
† CPF SKR	15mm	SP	57°,68°,79°,93°,141°,182° (5mm bulb)
† CPF SKR	15mm	SU	57°,68°,79°,93°,141°,182° (5mm bulb)
† CPF SKR	15mm	SU	57°, 68°, 79°, 93°, 141°,182° (8mm bulb)
† CPF SKR	15mm	SP	57°, 68°, 79°, 93°, 141°,182° (8mm bulb)

Globe Fire Sprinkler Corporation

4077 Air Park Drive, PO Box 796, Standish, Michigan 48658, USA

Tel: +1 517 846 4583 • Fax: +1 517 846 9231

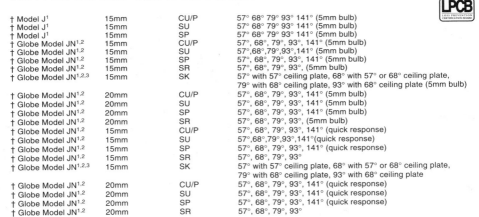

† Model J[1]	15mm	CU/P	57° 68° 79° 93° 141° (5mm bulb)
† Model J[1]	15mm	SU	57° 68° 79° 93° 141° (5mm bulb)
† Model J[1]	15mm	SP	57° 68° 79° 93° 141° (5mm bulb)
† Globe Model JN[1,2]	15mm	CU/P	57°, 68°, 79°, 93°, 141° (5mm bulb)
† Globe Model JN[1,2]	15mm	SU	57°,68°,79°,93°,141° (5mm bulb)
† Globe Model JN[1,2]	15mm	SP	57°, 68°, 79°, 93°, 141° (5mm bulb)
† Globe Model JN[1,2]	15mm	SR	57°, 68°, 79°, 93°, (5mm bulb)
† Globe Model JN[1,2,3]	15mm	SK	57° with 57° ceiling plate, 68° with 57° or 68° ceiling plate, 79° with 68° ceiling plate, 93° with 68° ceiling plate (5mm bulb)
† Globe Model JN[1,2]	20mm	CU/P	57°, 68°, 79°, 93°, 141° (5mm bulb)
† Globe Model JN[1,2]	20mm	SU	57°, 68°, 79°, 93°, 141° (5mm bulb)
† Globe Model JN[1,2]	20mm	SP	57°, 68°, 79°, 93°, 141° (5mm bulb)
† Globe Model JN[1,2]	20mm	SR	57°, 68°, 79°, 93°, (5mm bulb)
† Globe Model JN[1,2]	15mm	CU/P	57°, 68°, 79°, 93°, 141° (quick response)
† Globe Model JN[1,2]	15mm	SU	57°,68°,79°,93°,141°(quick response)
† Globe Model JN[1,2]	15mm	SP	57°, 68°, 79°, 93°, 141° (quick response)
† Globe Model JN[1,2]	15mm	SR	57°, 68°, 79°, 93°
† Globe Model JN[1,2,3]	15mm	SK	57° with 57° ceiling plate, 68° with 57° or 68° ceiling plate, 79° with 68° ceiling plate, 93° with 68° ceiling plate
† Globe Model JN[1,2]	20mm	CU/P	57°, 68°, 79°, 93°, 141° (quick response)
† Globe Model JN[1,2]	20mm	SU	57°, 68°, 79°, 93°, 141° (quick response)
† Globe Model JN[1,2]	20mm	SP	57°, 68°, 79°, 93°, 141° (quick response)
† Globe Model JN[1,2]	20mm	SR	57°, 68°, 79°, 93°

[1] Also approved with chromium finish
[2] Also approved with coloured finish
[3] Not to be installed in ceilings with a positive pressure in the void above.

Globe Sprinklers Europa S.A.

Avda. de las Flores, 13 Parque Empresarial "El Molino", 28970 Humanes de Madrid (Madrid), Spain

Tel: +34 1 606 3711 • Fax: +34 1 690 9561

† Globe Model JN[1]	15mm	CU/P	57°, 68°, 79°, 93°, 141° (5mm bulb)
† Globe Model JN[1]	15mm	SP	57°, 68°, 79°, 93°, 141° (5mm bulb)
† Globe Model JN[1]	15mm	SU	57°, 68°, 79°, 93°, 141° (5mm bulb)

[1] Also approved with chromium finish

Grinnell Corporation

2401 NE Loop 289, Municipal Drive, Lubbock, Texas 79403, USA

Tel: +1 806 765 6691 • Fax: +1 806 765 6765
Sales Enquiries: Tel: +31 5 328 3434 • Fax: +31 5 328 3377

LPCB
LOSS PREVENTION
CERTIFICATION BOARD

† GEM D	15mm	CU/P	260°
† GEM D	15mm	SU	260°
† GEM D	15mm	SP	260°
GEM F950[1]	15mm	CU/P	74° 100° 141° 177°
GEM F950[1]	15mm	SU	74° 100° 141° 177°
GEM F950[1]	15mm	SP	74° 100° 141° 177°
GEM F995[6]	15mm	SF	71° 100° (quick response)
GEM F996	15mm	WHF	71° 100°
GEM FR-1 (Nickel link)[6,7]	15mm	SP	74° (quick response)
GEM FR-1 (Nickel link)[6,7]	15mm	SU	74° (quick response)
† GEM F 976[3,4,5]	15mm	SK	57° with 57° ceiling plate, 68° with 57° ceiling plate, 79° with 74° ceiling plate, 93° with 74° ceiling plate (5mm bulb)
† GEM A [1,8]	10mm	SP	57° 68° 79° 93° 141° 182° (5mm bulb)
† GEM A [1,8]	10mm	SU	57° 68° 79° 93° 141° 182° (5mm bulb)
† GEM A [1,8]	15mm	SP	57° 68° 79° 93° 141° 182° (5mm bulb)
† GEM A [1,8]	15mm	SU	57° 68° 79° 93° 141° 182° (5mm bulb)
† GEM A [1,8]	15mm	CU/P	57° 68° 79° 93° 141° 182° (5mm bulb)
† GEM A [1,8]	15mm	WU/P	57° 68° 79° 93° 141° 182° (5mm bulb)
† GEM A	15mm	WH	57°,68°,79°,93°,141°,182° (5mm bulb)
† GEM A [1,8]	20mm	SP	57° 68° 79° 93° 141° 182° (5mm bulb)
† GEM A [1,8]	20mm	SU	57° 68° 79° 93° 141° 182° (5mm bulb)
† GEM A [1,8]	20mm	CU/P	57° 68° 79° 93° 141° 182° (5mm bulb)
† GEM A[1,6,8]	15mm	CU/P	57°, 68°, 79°, 93°, 141° (quick response).
† GEM A[1,6,8]	15mm	SU	57°, 68°, 79°, 93°, 141° (quick response)
† GEM A[1,6,8]	15mm	SP	57°, 68°, 79°, 93°, 141° (quick response)
† GEM A [1,6,8]	15mm	WU/P	57° 68° 79°, 93° 141° (quick response)
† GEM A	15mm	WH	57°,68°,79°,93°,141°
† GEM A[1,6,8]	20mm	CU/P	57°, 68°, 79°, 93°, 141° (quick response)
† GEM A[1,6,8]	20mm	SU	57°, 68°, 79°, 93°, 141° (quick response)
† GEM A[1,6,8]	20mm	SP	57°, 68°, 79°, 93°, 141° (quick response)
GEM ESFR-1	20mm	SP	74°, 101°

[1] Also approved with chromium finish.
[3] To be installed in accordance with specific installation instructions.
[4] Not to be installed in ceilings with a positive pressure in the void above. The ceiling plate is not to be repainted.
[5] Approved with chrome and coloured ceiling plates.
[6] The suitability of 'quick response' sprinklers for many risks is not yet proven.Use is therefore limited to where the purchaser, authority having jurisdiction and/or insurers consent to its use. Quick response sprinklers must be handled with care due to the fragile nature of the operating element, with packaging and instructions being marked accordingly.
[7] Every five years, all sprinklers should be visually examined, and representative samples removed and tested to the requirements of the authority and/or insurer.
[8] Also approved with polyester finish.

Grinnell Manufacturing (UK) Limited

Stockport Trading Estate, Yew Street, Stockport SK4 2JW

Tel: +44 (0)161 477 1886 • Fax: +44 (0)161 477 6729

† Grinnell Type A[1]	10mm	SP	57° 68° 79° 93° 141° 182°
† Grinnell Type A[1]	10mm	SU	57° 68° 79° 93° 141° 182°
† Grinnell Type A[1]	15mm	CU/P	57° 68° 79° 93° 141° 182°
† Grinnell Type A[1]	15mm	SP	57° 68° 79° 93° 141° 182°
† Grinnell Type A[1]	15mm	SU	57° 68° 79° 93° 141° 182°
† Grinnell Type A[1]	15mm	WU/P	57° 68° 79° 93° 141° 182°
† Grinnell Type AC[2]	15mm	SK	68° with coloured or chromium 57° ceiling plate.
† Grinnell Type A[1]	20mm	CU/P	57° 68° 79° 93° 141°
† Grinnell Type A[1]	20mm	SP	57° 68° 79° 93° 141°
† Grinnell Type A[1]	20mm	SU	57° 68° 79° 93° 141°
† Grinnell Type A[1]	10mm	SP	57°, 68°, 79°, 93°, 141°, 182°, (5mm bulb)
† Grinnell Type A[1]	10mm	SU	57°, 68°, 79°, 93°, 141°, 182°, (5mm bulb)
† Grinnell Type A[1]	15mm	SP	57°, 68°, 79°, 93°, 141°, 182°, (5mm bulb)
† Grinnell Type A[1]	15mm	SU	57°, 68°, 79°, 93°, 141°, 182°, (5mm bulb)
† Grinnell Type A[1]	15mm	CU/P	57°, 68°, 79°, 93°, 141°, 182°, (5mm bulb)
† Grinnell Type A[1,5]	15mm	SR	57°,68°,79°,93° (5mm bulb)
† Grinnell Type A[1]	15mm	WU/P	57°, 68°, 79°, 93°, 141°, 182°, (5mm bulb)
† Grinnell Type A	15mm	WH	57°,68°,79°,93°,141°,182° (5mm bulb)
† Grinnell Type A[1]	20mm	SP	57°, 68°, 79°, 93°, 141°, 182°, (5mm bulb)
† Grinnell Type A[1]	20mm	SU	57°, 68°, 79°, 93°, 141°, 182°, (5mm bulb)
† Grinnell Type A[1]	20mm	CU/P	57°, 68°, 79°, 93°, 141°, 182°, (5mm bulb)
† Grinnell Type A[1,5]	20mm	SR	57°,68°,79°,93° (5mm bulb)
† Grinnell Type A[1,3]	15mm	CU/P	57°, 68°, 79°, 93°, 141° (quick response)
† Grinnell Type A[1,3]	15mm	SU	57°, 68°, 79°, 93°, 141° (quick response)
† Grinnell Type A[1,3]	15mm	SP	57°, 68°, 79°, 93°, 141° (quick response)
† Grinnell Type A[1,5]	15mm	SR	57°,68°,79°,93°
† Grinnell Type A[1,3]	15mm	WU/P	57°, 68°, 79°, 93°, 141° (quick response)
† Grinnell Type A	15mm	WH	57°,68°,79°,93°,141°
† Grinnell Type A[1,5]	15mm	WHR	57°,68°,79°,93°
† Grinnell Type A[1,3]	20mm	CU/P	57°, 68°, 79°, 93°, 141° (quick response)
† Grinnell Type A[1,3]	20mm	SU	57°, 68°, 79°, 93°, 141° (quick response)
† Grinnell Type A[1,3]	20mm	SP	57°, 68°, 79°, 93°, 141° (quick response)
† Grinnell Type A[1,5]	20mm	SR	57°,68°,79°,93°
Grinnell ESFR-1	20mm	SP	74°, 101°
† Grinnell Type A[1,3]	10mm	SP	57°,68°,79°,93°,141° (quick response)
† Grinnell Type A[1,3]	10mm	SU	57°,68°,79°,93°,141° (quick response)

The sprinkler heads listed below are certificated with the EFSG Mark, LPCB Certificate No. 007a

† Grinnell Type A[1]	10mm	SP	57°, 68°, 79°, 93°, 141°, 182°, (5mm bulb)
† Grinnell Type A[1]	10mm	SU	57°, 68°, 79°, 93°, 141°, 182°, (5mm bulb)
† Grinnell Type A[1]	15mm	SP	57°, 68°, 79°, 93°, 141°, 182°, (5mm bulb)
† Grinnell Type A[1]	15mm	SU	57°, 68°, 79°, 93°, 141°, 182°, (5mm bulb)
† Grinnell Type A[1]	15mm	CU/P	57°, 68°, 79°, 93°, 141°, 182°, (5mm bulb)
† Grinnell Type A[1]	20mm	SP	57°, 68°, 79°, 93°, 141°, 182°, (5mm bulb)
† Grinnell Type A[1]	20mm	SU	57°, 68°, 79°, 93°, 141°, 182°, (5mm bulb)
† Grinnell Type A[1]	20mm	CU/P	57°, 68°, 79°, 93°, 141°, 182°, (5mm bulb)
† Grinnell Type A[1,3]	15mm	CU/P	57°, 68°, (quick response)
† Grinnell Type A[1,3]	15mm	SU	57°, 68°, (quick response)
† Grinnell Type A[1,3]	15mm	SP	57°, 68°, (quick response)
† Grinnell Type A[1,3]	20mm	CU/P	57°, 68°, (quick response)
† Grinnell Type A[1,3]	20mm	SU	57°, 68°, (quick response)
† Grinnell Type A[1,3]	20mm	SP	57°, 68°, (quick response)

[1]   Also approved with coloured and chromium finishes.
[2]   To be installed in accordance with specific installation instructions. Not to be installed in ceilings with a positive pressure in the void above. The ceiling plate is not to be repainted.
[3]   The suitability of 'quick response' sprinklers for many risks is not yet proven. Use is therefore limited to where the purchaser, authority having jurisdiction and/or insurers consent to its use. Quick response sprinklers must be handled with care due to the fragile nature of the operating element, with packaging and instructions being marked accordingly.
[5]   Approved with Model F700 recessed escutcheon

5 SECTION 11:
11.1 SPRINKLER HEADS

G.W. Sprinkler A/S.

Kastanievej 15, DK - 5620 Glamsbjerg, Denmark

Tel: +45 64 72 2055 • Fax: +45 64 72 2255

† GW-DD-F2 [1,5,6]	15mm	WHEC	57° 68° 79°
† GW-DD1[1]	10mm	SR	57°, 68°, 79°, 93° (5mm bulb)
† GW-DD1[1]	10mm	SP	57°, 68°, 79°, 93°, 141°, 182°, (5mm bulb)
† GW-DD1[1]	10mm	SU	57°, 68°, 79°, 93°, 141°, 182°, (5mm bulb)
† GW-DD1[1,2,7]	10mm	SR	57°, 68°, 79°, 93° (5mm bulb)
† GW-DD1[1,7]	10mm	SP	57°, 68°, 79°, 93°, 141°, 182°, (5mm bulb)
† GW-DD1[1,7]	10mm	SU	57°, 68°, 79°, 93°, 141°, 182°, (5mm bulb)
† GW-DD1[1]	15mm	CU/P	57°, 68°, 79°, 93°, 141°, 182°, (5mm bulb)
† GW-DD1[1,2]	15mm	SR	57°, 68°, 79°, 93° (5mm bulb)
† GW-DD1[1]	15mm	SP	57°, 68°, 79°, 93°, 141°, 182°, (5mm bulb)
† GW-DD1[1]	15mm	SU	57°, 68°, 79°, 93°, 141°, 182°, (5mm bulb)
† GW-DD1[1]	15mm	WU/P	57°, 68°, 79°, 93°, 141°, 182°, (5mm bulb)
† GW-DD1[1,3]	20mm	CU/P	57°, 68°, 79°, 93°, 141°, 182°, (5mm bulb)
† GW-DD1[1,3]	20mm	SP	57°, 68°, 79°, 93°, 141°, 182°, (5mm bulb)
† GW-DD1[1,3]	20mm	SU	57°, 68°, 79°, 93°, 141°, 182°, (5mm bulb)
† GW-DD1[1]	20mm	SP	57°, 68°, 79°, 93°, 141°, 182°, (5mm bulb)
† GW-DD1[1]	20mm	SU	57°, 68°, 79°, 93°, 141°, 182°, (5mm bulb)
† GW-DD1[1]	20mm	CU/P	57°, 68°, 79°, 93°, 141°, 182°, (5mm bulb)
† GW-DD1 QR[1,2,]	10mm	SR	57°, 68°, 79°, 93°, 141°,
† GW-DD1 QR[1,4]	10mm	SP	57°, 68°, 79°, 93°, 141°, (quick response)
† GW-DD1 QR[1,4]	10mm	SU	57°, 68°, 79°, 93°, 141°, (quick response)
† GW-DD1 QR[1,2,7]	10mm	SR	57°, 68°, 79°, 93°, 141°,
† GW-DD1 QR[1,4,7]	10mm	SP	57°, 68°, 79°, 93°, 141°, (quick response)
† GW-DD1 QR[1,4,7]	10mm	SU	57°, 68°, 79°, 93°, 141°, (quick response)
† GW-DD1 QR[1,2,]	15mm	SR	57°, 68°, 79°, 93°, 141°,
† GW-DD1 QR[1,4]	15mm	SP	57°, 68°, 79°, 93°, 141°, (quick response)
† GW-DD1 QR[1,4]	15mm	SU	57°, 68°, 79°, 93°, 141°, (quick response)
† GW-DD1 QR[1,4]	15mm	CU/P	57°, 68°, 79°, 93°, 141°, (quick response)
† GW-DD1 QR[1,4]	15mm	WU/P	57°, 68°, 79°, 93°, 141°, (quick response)
† GW-DD1 QR[1,3,4]	20mm	SP	57°, 68°, 79°, 93°, 141°, (quick response)
† GW-DD1 QR[1,3,4]	20mm	CU/P	57°, 68°, 79°, 93°, 141°, (quick response)
† GW-DD1 QR[1,3,4]	20mm	SU	57°, 68°, 79°, 93°, 141°, (quick response)
† GW-DD1 QR[1,4]	20mm	SP	57°, 68°, 79°, 93°, 141°, (quick response)
† GW-DD1 QR[1,4]	20mm	SU	57°, 68°, 79°, 93°, 141°, (quick response)
† GW-DD1 QR[1,4]	20mm	CU/P	57°, 68°, 79°, 93°, 141°, (quick response)

1 Also approved with chromium and painted finishes.
2 Approved with brass, chrome or coloured recessing cup.
3 This 20mm orifice sprinkler is fitted with a pintle to denote that it has a 15mm nominal thread size.
4 The suitability of 'quick response' sprinklers for many risks is not yet proven. Use is therefore limited to where the purchaser, authority having jurisdiction and/or insurers consent to its use. Quick response sprinklers must be handled with care due to the fragile nature of the operating element, with packaging and instructions being marked accordingly.
5 This extended coverage horizontal sidewall sprinkler should be handled with care to ensure that the deflector is not deformed. After installation the deflector angle should be checked with the manufacturers gauge to verify its angle of 92.5°.
6 The product manufacturer should be consulted to determine the product fire protection capability and installation practice. Available test evidence and the CEA document "Requirements and Test Methods for Hotel Room Horizontal Sidewall Sprinklers" should also be consulted to ensure correct usage in light hazard applications.
7 This sprinkler is fitted with a pintle to denote that it has a 15mm nominal thread size.

Minimax GmbH

Industriestrasse 10/12, D-23840 Bad Oldesloe, Germany

Tel: +49 4531 8030 • Fax: +49 4531 803 248

† SFH-Decor	15mm	WU	68° 79° 93° 141° 182°
† SFH-Decor	20mm	CU/P	68° 79° 93° 141° 182°
† SFH-Decor M[1]	15mm	CU/P	68° 79° 93° 141° 182°
† Type MXF[2]	15mm	CU/P	68° (quick response)
† Type MX[1]	15mm	CU/P	68°, 79°,93°,141°,182° (5mm bulb)
† Type MX[1]	15mm	SP	68°, 79°,93°,141°,182° (5mm bulb)
† Type MX[1]	15mm	SU	68°, 79°,93°,141°,182° (5mm bulb)

1 Approved with plain and chrome finishes.
2 The suitability of 'quick response' sprinklers for many risks is not yet proven. Use is therefore limited to where the purchaser, authority having jurisdiction and/or insurers consent to its use. Quick response sprinklers must be handled with care due to the fragile nature of the operation element, with packaging and instructions being marked accordingly.

The Reliable Automatic Sprinkler Co. Inc.

525, North MacQuesten Parkway, Mount Vernon, New York, 10552-2600 USA

Tel: +1 914 668 3470 • Fax: +1 914 668 2936
UK Sales Enquiries: Tel: +44 (0)1372 728899 • Fax: +44 (0)1372 724461

Reliable B [1,2]	15mm	SF	57°, 71°, 100°,
† Reliable F1 [2, 3]	10mm	SU	57°, 68°,79°, 93°, 141°, 182° (5mm bulb)
† Reliable F1 [2, 3, 4]	10mm	SP	57°, 68°,79°, 93°, 141°, 182° (5mm bulb)
† Reliable F1 [5, 13]	10mm	SR	57°, 68°, 79°, 93° (5mm bulb)
† Reliable F1 [20]	10mm	SU	57°, 68°, 79°, 93°, 141° (5mm bulb)
† Reliable F1 [20]	10mm	SP	57°, 68°, 79°, 93°, 141° (5mm bulb)
† Reliable F1 [20, 22, 24]	10mm	SR	57°, 68°, 79°, 93°, (5mm bulb)
† Reliable F1 [2, 3]	15mm	CU/P	57°, 68°,79°, 93°, 141°, 182° (5mm bulb)
† Reliable F1 [2, 3]	15mm	SU	57°, 68°,79°, 93°, 141°, 182° (5mm bulb)
† Reliable F1 [2, 3, 4]	15mm	SP	57°, 68°,79°, 93°, 141°, 182° (5mm bulb)
† Reliable F1 [2, 3]	15mm	WP	57°, 68°,79°, 93°, 141°, 182° (5mm bulb)
† Reliable F1 [5, 13]	15mm	SR	57°, 68°, 79° (5mm bulb)
† Reliable F1 [20]	15mm	SU	57°, 68°, 79°, 93°, 141° (5mm bulb)
† Reliable F1 [20]	15mm	SP	57°, 68°, 79°, 93°, 141° (5mm bulb)
† Reliable F1 [20]	15mm	WP	57°, 68°, 79°, 93°, 141° (5mm bulb)
† Reliable F1 [20, 22, 24]	15mm	SR	57°, 68°, 79°, 93° (5mm bulb)
† Reliable F1 [2,3]	20mm	CU/P	57°, 68°,79°,93°,141°, 182° (5mm bulb)
† Reliable F1	20mm	SU	57°, 68°,79°,93°,141°,182° (5mm bulb)
† Reliable F1	20mm	SP	57°, 68°,79°,93°,141°,182° (5mm bulb)
† Reliable F1 [20]	20mm	SU	57°, 68°, 79°, 93°, 141° (5mm bulb)
† Reliable F1 [20]	20mm	SP	57°, 68°, 79°, 93°, 141° (5mm bulb)
† Reliable F1 [22,24]	20mm	SR	57°, 68°, 79°, 93° (5mm bulb)
† Reliable F1FR [14, 20]	10mm	SU	57°, 68°, 79°, 93°, 141° (quick response)
† Reliable F1FR [14, 20]	10mm	SP	57°, 68°, 79°, 93°, 141° (quick response)
† Reliable F1FR [20, 21, 22]	10mm	SR	57°, 68°, 79°, 93°
† Reliable F1FR [20, 22, 23]	10mm	SR	57°, 68°, 79°, 93°
† Reliable F1FR [14, 20]	15mm	CU/P	57°, 68°, 79°, 93°, 141° (quick response)
† Reliable F1FR [14, 20]	15mm	SU	57°, 68°, 79°, 93°, 141° (quick response)
† Reliable F1FR [14, 20]	15mm	SP	57°, 68°, 79°, 93°, 141° (quick response)
† Reliable F1FR [14, 20]	15mm	WP	57°, 68°, 79°, 93°, 141° (quick response)
† Reliable F1FR [20, 21, 22]	15mm	SR	57°, 68°, 79°, 93°
† Reliable F1FR [20, 22, 23]	15mm	SR	57°, 68°, 79°, 93°
† Reliable F1FR [14, 20]	20mm	SU	57°, 68°, 79°, 93°, 141° (quick response)
† Reliable F1FR [14,20]	20mm	SP	57°, 68°, 79°, 93°, 141° (quick response)
† Reliable F1FR [22,23]	20mm	SR	57°, 68°,79°,93°
* Reliable G [2]	10mm	SU	57°, 74°, 100°, 141°
* Reliable G [2]	10mm	SP	57°, 74°, 100°, 141°
* Reliable G [2]	15mm	CU/P	57°, 74°, 100°, 141°
* Reliable G [2]	15mm	SU	57°, 74°, 100°, 141°
* Reliable G [2]	15mm	SP	57°, 74°, 100°, 141°
* Reliable G [2]	15mm	WU/P	57°, 74°, 100°, 141°
Reliable G [7,8,9]	15mm	SR	57°, 74°, 100°
* Reliable G [2]	20mm	CU/P	57°, 74°, 100°, 141°
* Reliable G [2]	20mm	SU	57,° 74°, 100°, 141°
* Reliable G [2]	20mm	SP	57°, 74°, 100°, 141°
Reliable G1 [9,10]	15mm	SK	(fixed and adjustable) 57° with 57° ceiling plate, 74° with 74° ceiling plate, 100° with 74° ceiling plate
Reliable G3 [2]	15mm	SDP	(fixed and adjustable) 57°, 74°
Reliable G/F [16,12,13]	10mm	SR	57°, 74° 100°
Reliable G/F [16,12,13]	15mm	SR	57°, 74° 100°
Reliable GFR [14, 16, 17]	10mm	SU	74° (quick response)
Reliable GFR [14, 16, 17]	10mm	SP	74° (quick response)
Reliable GFR [14 ,16, 17]	15mm	CU/P	74° (quick response)
Reliable GFR [14, 16, 17]	15mm	SU	74° (quick response)
Reliable GFR [14, 16, 17]	15mm	SP	74° (quick response)

Continued

181

5

SECTION 11:

11.1 SPRINKLER HEADS

The Reliable Automatic Sprinkler Co. Inc. (continued)

Reliable GFR[14, 16, 17]	15mm	WU/P	74° (quick response)
Reliable GFR/F1[16, 17, 18]	15mm	SR	74°
Reliable GFR/F2[16, 17, 19]	15mm	SR	74°
Reliable G4FR	15mm	SK	(adjustable ceiling plate) 74° with 57° ceiling plate
Reliable ZX - 1[2, 14, 16]	15mm	SF	74° (quick response)
† Reliable ESFR-H	20mm	SP	68°, 93°
†Reliable F1FR [14, 20]	20mm	CU/P	57°,68°,79°,93°,141° (quick response)

[1]   Not considered suitable for use in atmospheres which are corrosive or subject to high dust content. Approved with a white painted coverplate.

[2]   Approved with a chromium finish.

[3]   Also approved with plated brass, black plate and dull and bright chrome finishes.

[4]   The 57°, 68°, 79° and 93° sprinklers are also approved with a white paint finish.

[5]   Incorporating Reliable F1 SP Sprinkler in Model G/F1 recessed escutcheon. To be installed in accordance with installation bulletin 119.

[7]   To be installed in accordance with installation bulletin 111G. Approved with brass, chrome or coloured recessing cup.

[8]   The Reliable G has slotted holes in the recessing cup. (The Reliable G/F1, which is similar in appearance, has no slotted holes in the recessing cup.)

[9]   Not to be installed in ceilings with a positive pressure in the void above.

[10]   To be installed in accordance with installation bulletin 115. Approved with chrome and coloured ceiling plates. The ceiling plate shall not be repainted.

[12]   Incorporating Reliable G SP Sprinkler in Model G/F1 recessed escutcheon. To be installed in accordance with installation bulletin 120.

[13]   Approved with brass, chrome or white G/F1 recessed escutcheon.

[14]   The suitability of 'quick response' sprinklers for many risks is not yet proven.Use is therefore limited to where the purchaser, authority having jurisdiction and/or insurers consent to its use. Quick response sprinklers must be handled with care due to the fragile nature of the operating element, with packaging and instructions being marked accordingly.

[16]   Every five years, all sprinklers should be visually examined and representative samples removed and tested to the requirements of the authority and/or insurer.

[17]   Approved with natural, chrome and white painted finishes.

[18]   Incorporating Reliable GFR SP sprinkler in Model G/F1 recessed escutcheon.

[19]   Incorporating Reliable GFR SP sprinkler in Model F2 recessed escutcheon.

[20]   Approved with plated brass, bronze, chrome, black-plated and coloured finish.

[21]   Incorporating reliable F1FR SP sprinkler in Model F1 recessed escutcheon.

[22]   To be installed in accordance with specific installation instructions.

[23]   Incorporating Reliable F1FR SP sprinkler in Model F2 recessed escutcheon.

[24]   Incorporating Reliable F1 SP (5mm bulb) sprinkler in Model F1 recessed escutcheon.

[25]   Also approved with white painted cover and sensor and white cover with chrome sensor.

Rolland Sprinklers

Usine de 73410 Mognard, France

Tel: +33 479 547 250 • Fax: +33 479 547 248

† Rolland Model N[1]	15mm	SP	57°,68°,79°,93°,141° (quick response)
† Rolland Model N[1]	15mm	SU	57°,68°,79°,93°,141° (quick response)
† Rolland Model N[1]	15mm	CU/P	57°,68°,79°,93°,141° (quick response)
† Rolland Model N[1]	15mm	SP	57°,68°,79°,93°,141°,182° (5mm bulb)
† Rolland Model N[1]	15mm	SU	57°,68°,79°,93°,141°,182° (5mm bulb)
† Rolland Model N[1]	15mm	CU/P	57°,68°,79°,93°,141°,182° (5mm bulb)

[1] Also approved with chromium and painted finishes

Spraysafe Automatic Sprinklers Limited

Corringham Road Industrial Estate, Gainsborough, Lincolnshire DN21 1QB

Tel: +44 (0)1427 615401 • Fax: +44 (0)1427 610433

Spraysafe A [2, 3, 4]	15mm	SK	72° with 57° ceiling plate, 100° with 74° ceiling plate
† Spraysafe B[5]	10mm	SP	57° 68° 79° 93° 141°
† Spraysafe B[5]	10m[m]	SU	57° 68° 79° 93° 141°
† Spraysafe B[5]	15mm	CU/P	57° 68° 79° 93° 141°
† Spraysafe B[5]	15mm	SU	57° 68° 79° 93° 141°
† Spraysafe B[5]	15mm	SP	57° 68° 79° 93° 141°
† Spraysafe B[5]	15mm	WU/P	57° 68° 79° 93° 141°
† Spraysafe B[2, 3, 6, 7]	15mm	SR	57° 68°
† Spraysafe B[5]	20mm	CU/P	68° 79° 93° 141°
† Spraysafe BL020[8]	20mm	CU/P	68° 79°
† Spraysafe S10[5]	10mm	SP	57°,68°,79°,93°,141° (5mm bulb)
† Spraysafe S10[5]	10mm	SU	57°,68°,79°,93°,141° (5mm bulb)
† Spraysafe S15[5]	15mm	SP	57°,68°,79°,93°,141° (5mm bulb)
† Spraysafe S15[5]	15mm	SU	57°,68°,79°,93°,141° (5mm bulb)
† Spraysafe S15[5]	15mm	CU/P	57°,68°,79°,93°,141° (5mm bulb)
† Spraysafe S15[5]	15mm	SR	57°,68°,79° (5mm bulb)
† Spraysafe S15[5]	15mm	WU/P	57°,68°,79°,93°,141° (5mm bulb)
† Spraysafe SC10[5]	10mm	SU	68° (5mm bulb)(special response)
† Spraysafe SC10[5]	10mm	SP	68° (5mm bulb)(special response)
† Spraysafe SC15[5]	15mm	CU/P	68° (5mm bulb)(special response)
† Spraysafe SC15[5]	15mm	SU	68° (5mm bulb)(special response)
† Spraysafe SC15[5]	15mm	SP	68° (5mm bulb)(special response)
† Spraysafe SC15[5]	15mm	SR	68° (5mm bulb)
† Spraysafe SC15[5]	15mm	WU/P	68° (5mm bulb)(special response)
† Spraysafe SFR10[5,9]	10mm	SU	57°,68°,79°,93°,141° (quick response)
† Spraysafe SFR10[5,9]	10mm	SP	57°,68°,79°,93°,141° (quick response)
† Spraysafe SFR15[5,9]	15mm	CU/P	57°,68°,79°,93°,141° (quick response)
† Spraysafe SFR15[5,9]	15mm	SU	57°,68°,79°,93°,141° (quick response)
† Spraysafe SFR15[5,9]	15mm	SP	57°,68°,79°,93°,141° (quick response)
† Spraysafe SFR15[5]	15mm	SR	57°,68°,79° (3mm bulb)
† Spraysafe SFR15[5,9]	15mm	WU/P	57°,68°,79°,93°,141° (quick response)

2   To be installed with specific installation instructions.
3   Not to be installed in ceilings with a positive pressure in the void above.
4   Approved with chrome and coloured ceiling plates. The ceiling plate shall not be repainted.
5   Also approved with chrome and coloured finish.
6   To be installed in accordance with general arrangement drawing number 4-7150097
7   The escutcheon plate is approved in metal with plain, chrome, and coloured finish, and in thermosetting plastic with plain (i.e. black & white), chrome, and bronze finish.
8   This 20mm orifice sprinkler is fitted with a pintle to denote that it has a 1/2" nominal thread size.
9   The suitability of 'quick response' sprinklers for many risks is not yet proven. Use is therefore limited to where the purchaser, authority having jurisdiction and/or insurers consent to its use. Quick response sprinklers must be handled with care due to the fragile nature of the operating element, with packaging and instructions being marked accordingly.

Star Sprinkler Inc.

2401 N.E. Loop 289, Municipal Drive, Lubbock TX 79403, USA

Sales Enquiries: Tel: +1 414 570 5000 • Fax: +1 414 570 5010

Star LD2[1]	15mm	SP	57° 74° 100° 138° 183°
Star LD2 [3, 4, 5]	15mm	SR	57° 74° 100°
Star LD2 [1]	20mm	CU/P	57° 74° 100° 138° 183°
Star LD2 [1]	20mm	SU	57° 74° 100° 138° 183°
Star LD2 [1]	20mm	SP	57° 74° 100° 138° 183°
Star LD2 [3, 4, 5]	20mm	SR	57° 74° 100°
† Star SG	15mm	CDU/P	57°,68° (5mm bulb)
† Star SG	15mm	SDP	57°,68° (5mm bulb)
† Star SG[1, 6]	15mm	CU/P	57°, 68°,79°,93°,141°,182° (5mm bulb)
† Star SG[1, 6]	15mm	SU	57°, 68°,79°,93°,141°,182° (5mm bulb)
† Star SG[1, 6]	15mm	SP	57°, 68°,79°,93°,141°,182° (5mm bulb)
† Star SG[2, 3, 5]	15mm	SR	57° 68° 79° 93° (recessed and adjustable) (5mm bulb)
† Starmist SG [3, 4, 5]	15mm	SR	57° 68° 79° 93° (recessed and adjustable) (5mm bulb)
† Star SG[1, 6]	20mm	CU/P	57° 68° 79° 93° 141° 182° (5mm bulb)
† Star SG[1, 6]	20mm	SU	57° 68° 79° 93° 141° 182° (5mm bulb)
† Star SG[1, 6]	20mm	SP	57° 68° 79° 93° 141° 182° (5mm bulb)

Continued

Star Sprinkler Inc. (continued)

† Star SG[3, 5]	20mm	SR	57° 68° 79° 93° (recessed and adjustable) (5mm bulb)
† Star Q [3, 4, 7]	15mm	SK	57° with 57° ceiling plate, 68° with 57° ceiling plate, 79° with 74° ceiling plate, 93° with 74° ceiling plate (4mm bulb)
† Star SGQR [1, 3, 5]	15mm	SR	57° 68°
† Star SGQR [1, 6,]	15mm	WH	57° 68°
† Star SGQR [1, 6 ,8]	15mm	SU	57° 68° (quick response)
† Star SGQR [1, 6 ,8]	15mm	SP	57° 68° (quick response)
† Star SGQR [1,6, 8]	15mm	CU/P	57° 68° (quick response)
† Star SGQR [1,6, 8]	20mm	CU/P	57° 68° (quick response)
† Star SGQR [1,6, 8]	20mm	SU	57° 68° (quick response)
† Star SGQR [1,6, 8]	20mm	SP	57° 68° (quick response)
Star S120 (Nickel link)[8,9]	15mm	SU	74° (quick response)
Star S120 (Nickel link)[8,9]	15mm	SP	74° (quick response)
Star S120 (Nickel link)[8,9]	15mm	CU/P	74° (quick response)
Star S120 (Nickel link)[8,9]	20mm	CU/P	74° (quick response)

1   Also approved with a chromium finish.
3   To be installed in accordance with specific installation instructions.
4   Not to be installed in ceilings with a positive pressure in the void above.
5   Approved with brass, chrome or coloured recessing cup.
6   Also approved with a coloured finish.
7   Approved with brass, chrome or coloured ceiling plates.
8   The suitability of 'quick response' sprinklers for many risks is not yet proven. Use is therefore limited to where the purchaser, authority having jurisdiction and/or insurers consent to its use. Quick response sprinklers must be handled with care due to the fragile nature of the operating element, with packaging and instructions being marked accordingly.
9   Every five years, all sprinklers should be visually examined, and representative samples removed and tested to the requirements of the authority and/or insurer.

The Viking Corporation

210 N Industrial Park Road, Hastings, Michigan 49058, USA

Tel: +1 616 945 9501 8 • Fax: USA 616 945 9599

UK Sales Enquiries: Tel: +44 (0)1427 875999 • Fax: +44 (0)1427 875998

Viking H	15mm	SF	74° 104°
† Viking M [2,3]	10mm	SU	57°68°79°93°141°182°
† Viking M [2,3]	10mm	SP	57°68°79°93°141°182°
† Viking M [2,3]	15mm	CU/P	57°68°79°93°141°182°
† Viking M [2,3]	15mm	SU	57°68°79°93°141°182°
† Viking M [2,3]	15mm	SP	57°68°79°93°141°182°
† Viking M [2,3]	15mm	WH	57°68°79°93°141°182°
† Viking M [2,3]	20mm	CU/P	57°68°79°93°141°182°
† Viking M [2,3]	20mm	SU	57°68°79°93°141°182°
† Viking M [2,3]	20mm	SP	57°68°79°93°141°182°
† Viking M Microfast [4,5]	15mm	CU/P	57°68°79°93°141°(quick response)
† Viking M Microfast [4,5]	15mm	SU	57°68°79°93°141°(quick response)
† Viking M Microfast [4,5]	15mm	SP	57°68°79°93°141°(quick response)
† Viking F-1 Microfast [6]	15mm	SR	57°68°79°93°141°(recessed and adjustable)
Viking Model B-1[7, 13]	15mm	SK	72° sprinkler with 72° ceiling plate
† Viking M [11,12]	10mm	SU	57°68°79°93°141°182°(5mm bulb)
† Viking M [11,12]	10mm	SP	57°68°79°93°141°182°(5mm bulb)
† Viking M [11,12]	15mm	CU/P	57°68°79°93°141°182°(5mm bulb)
† Viking M [11,12]	15mm	SU	57°68°79°93°141°182°(5mm bulb)
† Viking M [11,12]	15mm	SP	57°68°79°93°141°182°(5mm bulb)
† Viking M [11,12]	15mm	WH	57°68°79°93°141°182°(5mm bulb)
† Viking M [11,12]	20mm	CU/P	57°68°79°93°141°182°(5mm bulb)
† Viking M [11,12]	20mm	SU	57°68°79°93°141°182°(5mm bulb)
† Viking M [11,12]	20mm	SP	57°68°79°93°141°182°(5mm bulb)
† Viking M [8]	15mm	SDP	57°68°79°93°(adjustable recessed)
† Viking M [9]	15mm	SDP	57°68°79°93°141°(adjustable standard)
† Viking M [10]	15mm	SDP	57°68°79°93°141°(plain barrel)

2   Approved with plain, chrome and polished brass finishes.
3   The 57°, 68°, 79° and 93° heads are approved with wax, and white polyester finish.
4   The suitability of 'quick response' sprinklers for many risks is not yet proven. Use is therefore limited to where the purchaser, authority having jurisdiction and/or insurers consent to its use. Quick response sprinklers must be handled with care due to the fragile nature of the operating element, with packaging and instructions being marked accordingly.
5   Approved with plain and chrome finishes.
6   Approved with brass, chrome or coloured recessing cup.
7   Approved with chrome, bright brass and painted finish coverplates.
8   Approved with chrome and white finishes.
9   Approved with chrome finish.
10 Approved with chrome and brass finish.
11 Approved with chrome and polished brass and coloured finishes.
12 The 57°, 68°, 79° and 93° heads are approved with wax finishes.
13 Not to be installed in ceilings with a positive pressure in the void above.

Viking S.A.

Zone Industrielle Haneboesch, L-4562 Differdange/Niedercorn, Luxembourg

Tel: +352 583737 • Fax: +352 583736
UK Sales Enquiries: Tel: +44 (0)1427 875999 • Fax: +44 (0)1427 87599

† Viking M [2,3]	10mm	SU	57°68°79°93°141°182°
† Viking M [2,3]	10mm	SP	57°68°79°93°141°182°
† Viking M [2,3]	15mm	CU/P	57°68°79°93°141°182°
† Viking M [2,3]	15mm	SU	57°68°79°93°141°182°
† Viking M [2,3]	15mm	SP	57°68°79°93°141°182°
† Viking M [2,3]	15mm	WH	57°68°79°93°141°182°
† Viking M [2,3]	20mm	CU/P	57°68°79°93°141°182°
† Viking M [2,3]	20mm	SU	57°68°79°93°141°182°
† Viking M [2,3]	20mm	SP	57°68°79°93°141°182°
† Viking M Microfast [4,5]	15mm	CU/P	57°68°(quick response)
† Viking M Microfast [4,5]	15mm	SU	57°68°(quick response)
† Viking M Microfast [4,5]	15mm	SP	57°68°(quick response)
† Viking F-1 Microfast [6]	15mm	SR	57°68°(recessed and adjustable)
†Viking M[2,3]	10mm	SU	57°,68°,79°,93°,141°,182° (5mm bulb)
†Viking M[2,3]	10mm	SP	57°,68°,79°,93°,141°,182° (5mm bulb)
†Viking M[2,3]	15mm	CU/P	57°,68°,79°,93°,141°,182° (5mm bulb)
†Viking M[2,3]	15mm	SU	57°,68°,79°,93°,141°,182° (5mm bulb)
†Viking M[2,3]	15mm	SP	57°,68°,79°,93°,141°,182° (5mm bulb)
†Viking M[2,3]	15mm	WH	57°,68°,79°,93°,141°,182° (5mm bulb)
†Viking M[2,3]	20mm	CU/P	57°,68°,79°,93°,141°,182° (5mm bulb)
†Viking M[2,3]	20mm	SU	57°,68°,79°,93°,141°,182° (5mm bulb)
†Viking M[2,3]	20mm	SP	57°,68°,79°,93°,141°,182° (5mm bulb)
†Viking M/E-1[6,7]	15mm	SR	57°,68°,79°,93°(5mm bulb)
†Viking M/E-1[6,7]	20mm	SR	57°,68°,79°,93°(5mm bulb)

[2]   Approved with plain, chrome, polished brass and coloured finish.
[3]   The 57°,68°,79° and 93° heads are approved with wax finishes.
[4]   The suitability of 'quick response' sprinklers for many risks is not yet proven. Use is therefore limited to where the purchaser, authority having jurisdiction and/or insurers consent to its use. Quick response sprinklers must be handled with care due to the fragile nature of the operating element, with packaging and instructions being marked accordingly.
[5]   Approved with plain and chrome finishes.
[6]   Approved with brass, chrome or coloured recessing cup.
[7]   Approved with recessed and adjustable escutcheon.

INTRODUCTION

Sprinkler Heads included in this Section include a means of operation in addition to the normal glass bulb mode.

Each entry includes the nominal orifice size in millimetres and the temperature rating in degrees centigrade.

Abbreviations used are as previously detailed in the LPCB List, Section 12.1 *Sprinkler Heads, Introduction.*

Approval Conditions

1. LPCB Certificated detectors compatible with the control and indicating equipment shall be used.

2. Suitable electrical detection control and indicating equipment compatible with the equipment below shall be used. The control and indicating equpment shall be LPCB Approved/Certificated.

3. Connecting cables shall comply with BS 6387:1994, classification C,W,Z.

4. The equipment must be installed, operated and maintained as prescribed in the manufacturer's Installation, Operating and Maintenance Manual.

5. To guard against inadvertent operation it is recommended that electrically operated sprinklers be used in conjunction with coincidence detection.

6. Where 'Metron' actuators are employed the maximum allowable operating temperature shall not be exceeded.

7. Where 'Metron' actuators are used, these shall also be LPCB approved.

8. The 'Guidance' in the LPCB List Section 23 *'Metron' Actuators,* shall also be taken into account.

GW Sprinkler A/S
Kastanievej 15, DK-5620 Glamsbjerg, Demmark

Tel: +45 64 722055 • Fax: +45 64 722256

† GW-DD1-EL[1,2,3]	10mm	SP	57°, 68°, 79° (5mm bulb)
† GW-DD1-EL[1,2,3]	10mm	SU	57°, 68°, 79° (5mm bulb)
† GW-DD1-EL[1,2,3,4]	10mm	SP	57°, 68°, 79° (5mm bulb)
† GW-DD1-EL[1,2,3,4]	10mm	SU	57°, 68°, 79° (5mm bulb)
† GW-DD1-EL[1,2,3]	15mm	CU/P	57°, 68°, 79° (5mm bulb)
† GW-DD1-EL[1,2,3]	15mm	SP	57°, 68°, 79° (5mm bulb)
† GW-DD1-EL[1,2,3]	15mm	SU	57°, 68°, 79° (5mm bulb)
† GW-DD1-QR EL[1,2,3]	10mm	SP	57°, 68°, 79° (quick response)
† GW-DD1-QR EL[1,2,3]	10mm	SU	57°, 68°, 79° (quick response)
† GW-DD1-QR EL[1,2,3,4]	10mm	SP	57°, 68°, 79° (quick response)
† GW-DD1-QR EL[1,2,3,4]	10mm	SU	57°, 68°, 79° (quick response)
† GW-DD1-QR EL[1,2,3]	15mm	SP	57°, 68°, 79° (quick response)
† GW-DD1-QR EL[1,2,3]	15mm	SU	57°, 68°, 79° (quick response)
† GW-DD1-QR EL[1,2,3]	15mm	CU/P	57°, 68°, 79° (quick response)

[1] Also approved with chrome and painted finishes.
[2] To be installed, operated and maintained in accordance with the GW Sprinkler A/S Installation, Operating and Maintenance Manual.
[3] Sprinklers marked EL incorporate a 'Metron' actuator.
[4] This sprinkler is fitted with a pintle to denote that it has a 15mm nominal thread size.

This Section lists Approved glass bulbs for sprinkler heads. The requirements for approval are that the sprinkler bulbs comply with the relevant requirements of LPS 1039:Issue 3 and are used in sprinkler heads listed in Section 12.

Day-Impex Limited
Station Works, Earls Colne, Colchester, Essex CO6 2ER

Tel: +44 (0)1787 223232 • Fax: +44 (0)1787 224171

Model	Bulb size (mm)	Temperature rating (°C)
941[1]	3	57, 68, 79, 93
937	5	57, 68, 79, 93, 141, 182
817	8	57, 68, 79, 93, 141

[1] Suitable for use in quick response sprinklers

Helmut Geissler Glasinstrumente GmbH
Leonhard-Karl-Str. 33, D-97877 Wertheim, Germany.

Tel: +49 9342/5057 • Fax: +49 9342/5035

Model	Bulb size (mm)	Temperature rating (°C)
E	5	57, 68, 79, 93, 141, 182

Job GmbH
Kurt-Fischer-Strasse 30, 22926 Ahrensburg, Hamburg, Germany

Tel: +49 4102 21140 • Fax: +49 4102 211470

Certificate No. 351a

Model	Bulb size (mm)	Temperature rating (°C)
F2.5 Quick Response[1]	2.5	68,93
F3 Quick Response[1]	3	57,68,79,93,141
G5	5	57,68,79,93,141,182

[1]Suitable for use in quick response sprinklers

Norbulb Sprinkler Elemente GmbH
Robert-Koch-Straße 25a, 22851 Norderstedt, Germany

Tel: +49 40 524 4081 • Fax: +49 40 524 4083

Certificate No. 428a

Model	Bulb size (mm)	Temperature rating (°C)
N5	5	57, 68, 79, 93, 141, 182

5

SECTION 13:

SUCTION TANKS FOR AUTOMATIC PUMPS

INTRODUCTION (including Approval Conditions)

Approval Conditions here set out, take precedence over any material contained in other documentation for the purposes of LPCB recognition.

1. The following proprietary tanks have been approved for use as pump suction tanks in conjunction with the LPC *Rules for Automatic Sprinkler Installations* and the LPCB *Conditions for Approval of Suction Tanks for Automatic Pumps*.

2. Clause 17.4.11.6 of the LPC *Rules for Automatic Sprinkler Installations* identifies the following types of pump suction tanks:

    Type A   Not dependent on inflow, fed from a potable water supply, and suitable for sprinkler service without emptying, cleaning, maintenance or repair for a period of not less than 15 years.

    Type B   Not dependent on inflow.

    Type C   Dependent on inflow.

    Type D   Dependent on inflow, fed from a potable water supply, and suitable for sprinkler service without emptying, cleaning, maintenance or repair for a period of not less than 15 years.

3. Each single-supply or multi-level take-off tank supplied under this approval scheme must not exceed a maximum capacity of 1300m³. Duplicate sub-divided tanks may have a total capacity not exceeding 2600m³, provided the dividing wall is structurally capable of supporting either full compartment whilst the other is empty, and the maximum capacity of each compartment does not exceed 1300m³.

4. Continuation of approval is conditional upon a current Water Research Centre (WRc) Certificate being held by the manufacturer, specific to each tank listed.

5.[1,2] **Galvanised steel tanks Type A and D shall have the tank shell protected by:**

    (a)   a coating weight of at least 610g/m² each side, with such tank entries marked 'g', or

    (b)   an equivalent corrosion protection system which has been specifically Approved and Listed by the LPCB, with such tank entries marked 'e'.

6.[1] Galvanised steel tanks Types B and C shall have the tank shell protected by a coating weight of at least 300g/m² each side, or an equivalent corrosion protection system, Approved and Listed by the LPCB.

7. Each tank approval entry specifically identifies the tank types which are LPCB approved as defined in para. 2 above (e.g. T99A, T99B, T99C and T99D).

8. The purchaser shall specify the required tank designation A, B, C or D in accordance with the LPC *Rules for Automatic Sprinkler Installations,* together with the tank reference number as shown in 7 above.

9. The tank must bear a plate stating:

    (i) The name and address of the manufacturer.

    (ii) The date of installation.

    (iii) The LPCB Reference Number, including the type suffix (A,B,C or D) which has been specified by the purchaser and agreed by the manufacturer.

    (iv) The maximum capacity in cubic metres. Multi-supply tanks must indicate the capacity for the automatic sprinkler system separately.

    Tanks which do not bear such a plate are not approved.

Note 1. Corrosion protection systems specified should be regarded as minimum levels and additional protection may be warranted in special circumstances to achieve a design life of at least 15 years (Types A & D) or 6 years (Types B & C).

Note 2. Where Type A or D tank shell exteriors are not protected by 610g/m^2 or equivalent, appropriate inspection, service and maintenance shall be specified. Exterior shell protection of 300g/m^2 galvanising is set as the minimum acceptable level of exterior protection, together with specified inspection service and maintenance.

Note 3. Immersion heaters used to prevent the ball valve and water in the vicinity from freezing shall be dual element type.

A.C.Plastic Industries Limited
Armstrong Road, Daneshill East, Basingstoke RG24 8NU

Tel: +44 (0)1256 29334 • Fax: +44 (0)1256 817862

	LPCB Ref. No.	Maximum capacity
Aquastow rectangular sectional GRP tank. Maximum height 3.5m.	T29A, T29B, T29C, T29D	1300m^3

Note: All tank approvals are strictly subject to tanks being supplied in accordance with LPCB Approval Conditions set out in the 'Introduction'. For galvanized steel tanks, Types A and D, the designation 'g' refers to tanks protected by 610g/m^2 each side and the designation 'e' refers to equivalent corrosion protection as set out in Item 5 of the Introduction.

Braithwaite Engineers Limited
Neptune Works, Newport, Gwent NP9 2UY

Tel: +44 (0)1633 262141 • Fax: +44 (0)1633 250631

	LPCB Ref. No.	Maximum capacity
Braithwaite, galvanised, rectangular, sectional, non-standard dimensions BS 1564 steel tank with galvanized steel roof.	T24A^g, T24B. T24C, T24D^g	1300m^3
Braithwaite (Bridgestone) rectangular, sectional GRP tank. Maximum height 4m.	T32A, T32B, T32C, T32D	500m^3

Note: All tank approvals are strictly subject to tanks being supplied in accordance with LPCB Approval Conditions set out in the 'Introduction'. For galvanized steel tanks, Types A and D, the designation 'g' refers to tanks protected by 610g/m^2 each side and the designation 'e' refers to equivalent corrosion protection as set out in Item 5 of the Introduction.

5 SECTION 13:
SUCTION TANKS FOR AUTOMATIC PUMPS

Dewey Waters Limited
Cox's Green, Wrighton, Bristol BS18 7QS

Tel: +44 (0)1934 862601 • Fax: +44 (0)1934 862602

	LPCB Ref. No.	Maximum capacity
Rectangular, sectional, contact moulded GRP tank. Maximum height 3m. Approved for use where not exposed to sunlight.	T33A, T33B, T33C, T33D	75m³
Rectangular, sectional (1200 module), contact moulded GRP tank. Maximum height 2.4m.	T34A, T34B, T34C, T34D	300m³

Note: All tank approvals are strictly subject to tanks being supplied in accordance with LPCB Approval Conditions set out in the 'Introduction'. For galvanized steel tanks, Types A and D, the designation 'g' refers to tanks protected by 610g/m² each side and the designation 'e' refers to equivalent corrosion protection as set out in Item 5 of the Introduction.

Franklin Hodge Industries Limited
Ramsden Road, Rotherwas Industrial Estate, Hereford HR2 6LR

Tel: +44 (0)1432 269605 • Fax: +44 (0)1432 277454

	LPCB Ref. No.	Maximum capacity
Bolted, aluminium panel tank lined internally with butyl rubber, with aluminium deck roof:		
- Cylindrical	T12A, T12B, T12C, T12D	1300m³
- Rectangular	T15A, T15B, T15C, T15D	1300m³
Bolted, galvanised steel panel tank lined internally with butyl rubber:		
- Cylindrical, with galvanised steel decking, externally coated with PVC; or	T16Aᵉ, T16B, T16C, T16Dᵉ	1300m³
- Rectangular, with aluminium deck roof.	T17Aᵉ, T17B, T17C, T17Dᵉ	1300m³
Bolted, cylindrical, galvanised steel plate tank,with the walls internally coated with bitumen paint in accordance with BS 3416, sealed to a concrete base, with one of the following roofs: (i) galvanised steel decking, externally coated with PVC or; (ii) aluminium decking.	T25Aᵍ, T25B, T25C, T25Dᵍ	1300m³

Note: All tank approvals are strictly subject to tanks being supplied in accordance with LPCB Approval Conditions set out in the 'Introduction'. For galvanized steel tanks, Types A and D, the designation 'g' refers to tanks protected by 610g/m² each side and the designation 'e' refers to equivalent corrosion protection as set out in Item 5 of the Introduction.

Galglass Limited
321 Hough Lane, Wombwell, Barnsley, South Yorkshire S73 0LR

Tel: +44 (0)1226 340370 • Fax: +44 (0)1226 756170

	LPCB Ref. No.	Maximum capacity
Bolted, cylindrical, galvanised steel plate tank lined internally with bitumen paint in accordance with BS 3416, with a galvanised steel decking roof externally coated with PVC.	T21Aᵍ, T21B, T21C, T21Dᵍ	1300m³
Bolted, cylindrical, vitreous enamelled steel plate tank, with a galvanised steel decking roof externally coated with PVC.	T23A, T23B, T23C, T23D	1300m³

Note: All tank approvals are strictly subject to tanks being supplied in accordance with LPCB Approval Conditions set out in the 'Introduction'. For galvanized steel tanks, Types A and D, the designation 'g' refers to tanks protected by 610g/m² each side and the designation 'e' refers to equivalent corrosion protection as set out in Item 5 of the Introduction.

Irish Industrial Tanks Limited

Unit C1, Ballymount Drive Industrial Estate, Ballymount Road, Walkinstown, Dublin 12, Ireland

Tel: +353 1 450 7893 • Fax: +353 1 450 6404

LPCB

	LPCB Ref. No.	Maximum capacity
Bolted, cylindrical, glass-lined, mild steel storage tank with a galvanized steel decking roof externally coated with PVC.	T28A, T28B, T28C, T28D	1300m³

Note: All tank approvals are strictly subject to tanks being supplied in accordance with LPCB Approval Conditions set out in the 'Introduction'. For galvanized steel tanks, Types A and D, the designation 'g' refers to tanks protected by 610g/m² each side and the designation 'e' refers to equivalent corrosion protection as set out in Item 5 of the Introduction.

Lancaster Bros Limited

Elmsfield Park, Holme, Carnforth, Lancashire LA6 1RJ

Tel: +44 (0)1539 563421 • Fax: +44 (0)1539 563949

LPCB

	LPCB Ref. No.	Maximum capacity
Bolted, cylindrical, galvanised steel plate tank lined internally with bitumen paint in accordance with BS 3416, with a galvanised steel decking roof externally coated with PVC.	T31A^g, T31B, T31C, T31D^g	1300m³
Bolted, cylindrical, vitreous enamelled steel storage tank, with a galvanized steel decking roof externally coated with PVC.	T36A, T36B, T36C, T36D	1300m³

Note: All tank approvals are strictly subject to tanks being supplied in accordance with LPCB Approval Conditions set out in the 'Introduction'. For galvanized steel tanks, Types A and D, the designation 'g' refers to tanks protected by 610g/m² each side and the designation 'e' refers to equivalent corrosion protection as set out in Item 5 of the Introduction.

Sekisui Plant Systems Co Limited

(Incorporating Shizuoka Sekisui Panel Tank Company Limited)
77 Kami-Onogo, Iwata City, Shizuoka Prefecture 438, Japan

Tel: +81 06 440 2508 • Fax: +81 06 440 2518

LPCB

	LPCB Ref. No.	Maximum capacity
Sekisui rectangular, sectional, hot press moulded panel, GRP tank. Maximum height 4m.	T35A, T35B, T35C, T35D	1000m³

Note: All tank approvals are strictly subject to tanks being supplied in accordance with LPCB Approval Conditions set out in the 'Introduction'. For galvanized steel tanks, Types A and D, the designation 'g' refers to tanks protected by 610g/m² each side and the designation 'e' refers to equivalent corrosion protection as set out in Item 5 of the Introduction.

Vulcan Tanks Limited

Cotes Park Lane, Cotes Park Industrial Estate, Alfreton, Derbyshire DE55 4NJ

Tel: +44 (0)1773 835321 • Fax: +44 (0)1773 836578

	LPCB Ref. No.	Maximum capacity
Bolted, cylindrical, galvanised steel plate tank internally sprayed with a bitumen solution in accordance with BS 3416, with a galvanised steel decking roof externally coated with PVC.	T14A^g, T14B, T14C, T14D^g	1300m³
Bolted, cylindrical, vitreous enamelled steel plate tank, with a galvanized steel decking roof externally coated with PVC.	T20A, T20B, T20C, T20D	825m³
Bolted, rectangular, 2.5m high, galvanised steel panel tank, with the walls internally coated with bitumen paint in accordance with BS 3416, sealed to a concrete base, with a galvanized steel decking roof externally coated with PVC.	T30A^g, T30B, T30C, T30D^g	1300m³

Note: All tank approvals are strictly subject to tanks being supplied in accordance with LPCB Approval Conditions set out in the 'Introduction'. For galvanized steel tanks, Types A and D, the designation 'g' refers to tanks protected by 610g/m² each side and the designation 'e' refers to equivalent corrosion protection as set out in Item 5 of the Introduction.

SECTION 14:
VORTEX INHIBITORS

The following inhibitors have been approved for use under the LPC *Rules for automatic sprinkler installations.*

Franklin Hodge Industries Limited
Rotherwas Industrial Estate, Hereford HR2 6LR

Tel: +44 (0)1432 269605 • Fax: +44 (0)1432 277454

F.H.I. Anti - Vortex Plate suitable for the following nominal bore suction pipes (mm):
100, 150, 200, 250, 300, 350, 400 and 450

Grinnell Manufacturing (UK) Limited
Stockport Trading Estate, Yew Street, Stockport SK4 2JW

Tel: +44 (0)161 477 1886 • Fax: +44 (0)161 477 6729

Inhibitors suitable for the following nominal bore suction pipes (mm):
65, 80, 100, 150, 200, 250, 300, 350 and 400.

Lancaster Bros. Limited
Elmsfield Park, Holme, Camforth, Lancashire LA6 1RJ

Tel: +44 (0)1539 563421 • Fax: +44 (0)1539 563949

Inhibitors suitable for the following nominal bore suction pipes (mm):
100,150,200,250,300,350,400,450

Vulcan Tanks Limited
Cotes Park Lane, Cotes Park Industrial Estate, Alfreton Derbyshire DE55 4NJ

Tel: +44 (0)1773 835321 • Fax: +44 (0)1773 836578

Inhibitors suitable for the following nominal bore suction pipes (mm):
150, 200, 250 and 300.

Potter Electric Signal Co.

2081 Craig Road, St Louis, Missouri 63146, USA

Tel: +1 314 878 4321 • Fax: +1 314 878 7264

| Potter Type VSR-F | 50mm, 65mm, 75mm, 100mm, 125mm, 150mm, 200mm |
| | (2-in, 2.5-in, 3-in, 4-in, 5-in, 6-in, 8-in, sizes) |

System Sensor (A Division of Pittway)

3825 Ohio Avenue, Saint Charles, Illinois 60174, USA

Tel: +1 630 377 6580 • Fax: +1 630 377 6645

European Sales Enquiries: System Sensor, Pittway Tecnologica SpA, Via Caboto, 19, 34147, Trieste, Italy

Tel: +39 40 9490 111 • Fax: +39 40 382 137

TYPE WDF	50mm, 65mm, 75mm, 100mm, 125mm, 150mm, 200mm. (Maximum working pressure 15 bar.)
WFD20E	50mm	(Max. working pressure 17.25 bar)
WFD25E	65mm	(Max. working pressure 17.25 bar)
WFD30E	80mm	(Max. working pressure 17.25 bar)
WFD40E	100mm	(Max. working pressure 17.25 bar)
WFD60E	150mm	(Max. working pressure 17.25 bar)
WFD80E	200mm	(Max. working pressure 17.25 bar)

SECTION 15:2 5
FLOW SWITCH TESTING DEVICES

Project Fire Engineers Limited

Sandyford Street, Stafford ST16 3NF

Tel: +44 (0)1785 222 999 • Fax: +44 (0)1785 222 959

"ZONECHECK" flow switch automatic tester system

Guidance and approval conditions

Zonecheck shall be installed in accordance with Project Fire Engineers Limited 'Instruction Booklet', but subject to guidance below taking precedence.

1. The Zonecheck system circulates, but does not draw-off water from the sprinkler system and consequently does not test any zone or control valves.
2. System approval based on use with LPCB approved and listed flow switch.
3. Effective operation of the Zonecheck system should be monitored.
4. In the event of any servicing being warranted on this system, it shall be ensured that this does not impair effective sprinkler protection.
5. Provision to test flow switches in the normal way (LPC Sprinkler Rules) should also be incorporated.
6. It shall be ensured that the associated pipework is adequately vented and charged hydraulically.
7. The automatic means of operation and recording does not form part of this approval.

5 SECTION 16: WATER SPRAY SYSTEMS
16.1 WATER SPRAY SYSTEMS (MEDIUM AND HIGH VELOCITY)

The following systems are approved under the LPC *Rules for automatic sprinkler installations* for use on oil and flammable liquid hazards.

Angus Fire Armour Limited
Kastanievej 15, DK-5620, Glamsbjerg, Denmark

UK Sales Enquiries: Tel: +44 (0)184 421 4545 • Fax: +44 (0)184 421 3511

Fyrhed High Velocity Waterspray System
Thermospray Medium Velocity Waterspray System

Atlas Fire Engineering Limited
Unit House, 2/8 Morfa Road, Swansea SA1 2HS

Tel: +44 (0)1792 465006 • Fax: +44 (0)1792 648535

Oilfyre System
Titan Fyrejet System
Titan Firespray System

Grinnell Manufacturing (UK) Limited
Stockport Trading Estate, Yew Street, Stockport SK4 2JW

Tel: +44 (0)161 477 1886 • Fax: +44 (0)161 477 6729

Mulsifyre System
Protectospray System

The Reliable Automatic Sprinkler Co. Inc.
525 North MacQuesten Parkway, Mount Vernon, New York, 10552-2600 USA

Tel: +1 914 668 3470 • Fax: +1 914 668 2936
UK Sales Enquiries: Tel: +44 (0)1372 728899 • Fax: +44 (0)1372 724461

Reliable Medium Velocity Water Spray System

Star Sprinkler Inc.
2401 N.E. Loop 289, Municipal Drive, Lubbock TX 79403, USA

Sales Enquiries:
Tel: +1 414 570 5000 • Fax: +1 414 570 5010

Star Medium Velocity Water Spray System

Total Walther GmbH,

Feuerschutz und Sicherheit, Waltherstr 51, 51069 Kôln-Dellbruck, Germany

Tel: +49 221 67 85 631 • Fax: +49 221 67 85 727

Walther High Velocity Spray System

The Viking Corporation

210 N Industrial Park Road, Hastings, Michigan 49058, USA

Tel: +1 616 945 9501 • Fax: +1 616 945 9599
Sales Enquiries: Tel: +352 583737 • Fax: +352 583736

Viking Medium Velocity Water Spray System
Viking High Velocity Water Spray System

Wormald Ansul (UK) Limited, Wormald Fire Systems

Wormald Park, Grimshaw Lane, Newton Heath, Manchester M40 2WL

Tel: +44 (0)161 205 2321 • Fax: +44 (0)161 455 4459

Mulsifyre System
Protectospray System
Wormald High Velocity Water Spray System
Wormald Medium Velocity Water Spray System

Spraysafe Automatic Sprinklers Limited
Corringham Road Industrial Estate, Gainsborough, Lincolnshire DN21 1QB

Tel: +44 (0)1427 615401 • Fax: +44 (0)1427 610433

Delspray Medium Velocity Sprayers

Model	Nominal orifice size (mm)	Cone angles	Material
A	6.5	95° 100° 110° 120°	Brass
B	7.5	95° 100° 110° 120°	Brass
C	8.5	95° 100° 110° 120°	Brass
D	9.5	95° 100° 110° 120°	Brass
E	11.0	95° 100° 110° 120°	Brass

The Spraysafe Delspray Medium Velocity Sprayers are approved with plain brass, nickel, lead and polyester finishes.

Angus Fire Armour Limited

Kastanievej 15, DK-5620, Glamsbjerg, Denmark

UK Sales Enquiries: Tel: +44 (0)1844 214545 • Fax: +44 (0)1844 213511

Angus R [1,2,3,] 15mm long-throw horizontal open sprinkler

[1] Also approved with chromium finish.
[2] To be installed in accordance with the supplier's installation instructions.
[3] Application is limited to where purchaser, authority having jurisdiction and/or insurer consent to use.

Central Sprinkler Company

Corringham Road Industrial Estate, Gainsborough, Lincolnshire DN21 1QB

Tel: +44 (0)1427 615401 • Fax: +44 (0)1427 610433
Sales Enquiries: Tel: +1 215 362 0700 • Fax: +1 215 362 5385

Central Model SS07 Open Sprinkler [1,2]

[1] To be used in accordance with the manufacturer's recommendations.
[2] Application is limited to where purchaser, authority having jurisdiction, and/or insurer consent to use.

Spraysafe Automatic Sprinklers Limited

Corringham Road Industrial Estate, Gainsborough, Lincolnshire DN21 1QB

Tel: +44 (0)1427 615401 • Fax: +44 (0)1427 610433

Spraysafe Model SS07 Open Sprinkler[1,2]

[1] To be used in accordance with the manufacturer's recommendations.
[2] Application is limited to where purchaser, authority having jurisdiction, and/or insurer consent to use.

SECTION 17:
17.1 FIRE PUMPS

INTRODUCTION

The following fire pumps are LPCB approved to LPS 1131/G - *Pumps for Automatic Sprinkler Installation Pump Sets* - for use in accordance with the *LPC Rules for automatic sprinkler installations* or equivalent standard.

Firms listed in this Section shall control and be responsible for the design, construction, testing and performance of all fire pump sets incorporating their LPCB approved fire pumps.

All fire pump sets incorporating LPCB approved fire pumps shall conform to the *LPC Rules for Automatic Sprinkler Installations,* or Rules specified in the contract, and shall be tested in accordance with BS 5306 : Part 2, clause 17.4.13.11.

Note: Only LPCB approved fire pumps shall be used to comply with the *LPC Rules for automatic sprinkler installations.*

Pumps are approved for a maximum working pressure of 12 bar except where stated.

Armstrong Pumps Limited
Peartree Road, Stanway, Colchester, Essex CO3 5JX

Tel: +44 (0)1206 579491 • Fax: +44 (0)1206 760532

Certificate No. 197a

Model	Type	Suction Inlet Diameter/Discharge Outlet Diameter (mm)	Impeller Diameter (mm)	Rated Speed (rev/min)	Rated Flow (L/min)(at NPSH req. = 5.38m)	Rated Head (bar)(at NPSH req. ≤ 5.38m)	Closed Valve Head (bar)
Series	Direct	200/125	292	2900	7450	7.5	10
4030 FP	coupled		241	2900	7000	3.8	6.8
8 x 5 x	end suction		292	2900)*	7450	7.5	10
11½			292	1450)*	4800	1.6	2.3
Series	Close	200/125	292	2900	7450	7.5	10
4020 FP	coupled		241	2900	7000	3.8	6.8
8 x 5 x	end suction		292	2900)*	7450	7.5	10
11½			292	1450)*	4800	1.6	2.3
4030FP	Direct coupled	150/100	259)*	2900	7680	5.3	8.6
6x4x10	end suction		205)*	2900	5280	3.6	4.8
			259	2900)*	7680	5.3	8.6
			259	1450)	3120	1.6	2.1
4020FP	Close coupled	150/100	259)*	2900	7680	5.3	8.6
6x4x10	end suction		205)*	2900	5280	3.6	4.8
			259	2900)*	7680	5.3	8.6
			259	1450)*	3120	1.6	2.1
4030FP[1]	Direct coupled	100/80	224	2900	3360	3.4	6.3
4x3x8½	end suction		175	2900	2460	1.9	3.9
4020FP[1]	Close coupled	100/80	224	2900	3360	3.4	6.3
4x3x8½	end suction		175	2900	2460	1.9	3.9
4030FP	Direct coupled	80/50	157	2900	1300	1.2	2.9
3x2x6	end suction		113	2900	750	0.9	1.5
4020FP	Close coupled	80/50	157	2900	1300	1.2	2.9
3x2x6	end suction		113	2900	750	0.9	1.5

)*These pumps are approved with impeller diameters and, where applicable, speeds within the ranges shown, to provide flows and heads within those ranges.
[1] Maximum working pressure: 10 bar.
The above pumps are fitted with 'closed head' relief valve Part No. UKDO 5493 - 145 Issue A.

H.J. Godwin Limited
Quenington, Cirencester, Gloucestershire GL7 5BX

Tel: +44 (0)1285 750271 • Fax: +44 (0)1285 750352

Certificate No. 224a

Model	Type	Suction Inlet Diameter/ Discharge Outlet Diameter (mm)	Impeller Diameter (mm)	Rated Speed (rev/min)	Rated Flow (L/min)(at NPSH req. ≤5.38m)	Rated Head (bar)(at NPSH req. ≤5.38m)	Closed Valve Head (bar)
Isoflow Model 100-80-160	Direct coupled end suction	100/80	174 140	2985 2985	3150 1800	0.7 0.9	4.1 2.7
Isoflow Unibloc Model 100-80-160	Close coupled end suction	100/80	174 140	2985 2985	3150 1800	0.7 0.9	4.1 2.7
Isoflow Model 100-65-200	Direct coupled end suction	100/65	214 165	2985 2985	2400 2250	2.8 0.8	6.5 3.8
Isoflow Unibloc Model 100-65-200	Close coupled end suction	100/65	214 165	2985 2985	2400 2250	2.8 0.8	6.5 3.8
Isoflow Model 80-50-200	Direct coupled end suction	80/50	214 165	2985 2985	1580 1160	1.3 1.0	6.1 3.5
Isoflow Unibloc Model 80-50-200	Close coupled end suction	80/50	214 165	2985 2985	1580 1160	1.3 1.0	6.1 3.5
Isoflow Model 125-100-260	Direct coupled end suction	125/100	264 210	2985 2985	5500 5250	7.1 2.5	10.3 6.1
Isoflow Model 50-32-200	Direct coupled end suction	50/32	214 165	2985 2985	385 275	3.9 2.3	6.2 3.7
Isoflow Unibloc Model 50-32-200	Close coupled end suction	50/32	214 165	2985 2985	385 275	3.9 2.3	6.2 3.7
Isoflow Model 125-80-200	Direct Coupled end suction	125/80	214 165	2990 2990	4285 3485	2.8 1.0	6.5 3.8
Isoflow Unibloc Model 125-80-200	Close coupled end suction	125/80	214 165	2990 2990	4285 3485	2.8 1.0	6.5 3.8

These pumps are approved with impeller diameters within the ranges shown, to provide flows and heads within those ranges.

The above pumps are fitted with 'closed head' relief valve Part No. 44-0474-9912 (The above pumps are approved with mechanical seals).

SECTION 17:
17.1 FIRE PUMPS

KSB Ajax Pumps Pty. Limited
27 Indwe Street, Tottenham, Victoria 3012, Australia

Tel: +61 3 314 0611 • Fax: +61 3 314 7435

Certificate No. 198a

Model	Type	Suction Inlet Diameter/ Discharge Outlet Diameter (mm)	Impeller Diameter (mm)	Rated Speed (rev/min)	Rated Flow (L/min)(at NPSH req. ≤ 5.38m)	Rated Head (bar)(at NPSH req. ≤ 5.38m)	Closed Valve Head (bar)
E100-26	Direct coupled end suction	125/100	230}*	2950	4700	5.9	7.5
			205}*	2950	3800	5.0	5.9
			230	2950}*	4700	5.9	7.5
			230	1800}*	3000¹	2.1	2.8
E100-20	Direct coupled end suction	125/100	214}*	2950	4500	5.5	6.4
			175}*	2950	3000	3.5	4.2
			214	2950}*	4500	5.5	6.4
			214	1800}*	2700¹	2.0¹	2.4
E80-20	Direct coupled end suction	100/80	214}*	2950	3800	5.4	6.5
			165}*	2950	2800	2.5	3.7
			214	2950}*	3800	5.4	6.5
			214	1800}*	2350¹	1.9	2.4
E65-20	Direct coupled end suction	80/65	195}*	2950	2500	3.5	5.4
			165}*	2950	2100	2.3	3.8
			195	2950}*	2500	3.5	5.4
			195	1800}*	1800	1.0	2.0
E65-16	Direct coupled end suction	80/65	174}*	2930	2250	2.6	4.1
			125}*	2930	1450	1.2	2.0
			174	2930}*	2250	2.6	4.1
			174	1800}*	1500¹	0.9	1.6
E50-20	Direct coupled end suction	65/50	214}*	2930	1350	4.6	6.4
			165}*	2930	1000	2.4	3.8
			214	2930}*	1350	4.6	6.4
			214	1800}*	850	1.6	2.4
E50-16	Direct coupled end suction	65/50	174}*	2930	1500	3.1	4.4
			135}*	2930	1000¹	1.6	2.5
			174	2930}*	1500	3.1	4.4
			174	1800}*	870¹	1.4	1.7

}* These pumps are approved with impeller diameters and speeds within the ranges shown, to provide flows and heads within those ranges.

The above pumps are fitted with 'closed head' relief valve Part no. 5-70.1.

(The above pumps are approved with packed glands or mechanical seals.)

SPP Pumps

Sterling Fluid Systems Limited

Theale Cross, Reading, Berkshire RG31 7SP

(Also at Crucible Close, Coleford, Gloucester GL16 8PS)

Tel: +44 (0)1189 323123 • Fax: +44 (0)1189 323302

Certificate No. 111b

Model	Type	Suction Inlet Diameter/ Discharge Outlet Diameter (mm)	Impeller Diameter (mm)	Rated Speed (rev/min)	Rated Flow (L/min)(at NPSH req. = 5.38m)	Rated Head (bar)(at NPSH req. = 5.38m)	Closed Valve Head (bar)
TD20D[2]	Horizontal	250/200	340)*	2950	13600	10.8	15.7[1]
Thru-stream	split case		275)*	2950	12000	7.6	10.3[1]
			273)*	2950	11600	7.4	10.0
			245)*	2950	8500	6.0	8.0
			340	2950)*	13600	10.8	15.7[1]
			340	2400)*	13500	6.0	10.3[1]
			340	2200)*	13300	4.4	8.8
TD 20E[2]	Horizontal	250/200	290)*	2950	13300	6.7	11.3[1]
Thru-stream	split case		273)*	2950	13300	5.7	10.0
			240)*	2950	10800	4.9	7.7
			290	2950)*	13300	6.7	11.3[1]
			290	2800)*	13800	5.3	10.2
			290	2200)*	12700	2.0	6.3
TD15F[3]	Horizontal	200/150	325)*	2950	9600	7.9	13.3[1]
Thru-stream	split case		282)*	2950	9600	5.4	10.0
			275)*	2950	9600	5.1	9.5
			245)*	2950	9600	3.9	7.5
			325	2950)*	9600	7.9	13.3[1]
			325	2550)*	9600	5.0	10.0
			325	2200)*	9600	2.7	7.3
KP15Y	Direct coupled	200/150	272)*	2950	7800	7.8	9.8
Unistream	end suction		239)*	2950	7600	5.2	7.7
			272	2950)*	7800	7.8	9.8
			272	2200)*	8200	3.2	5.6
TD15A[3]	Horizontal	200/150	325)*	2950	7300	12.8	14.4[1]
Thru-stream	split case		270)*	2950	7300	8.5	9.9
			240)*	2950	7300	5.9	7.7
			270	2950)*	7300	8.5	9.8
			270	2400)*	8650	4.3	6.4
			325	2950)*	7300	12.8	14.4[1]
			325	2450)*	8200	7.7	10.2
			325	2200)*	8700	6.0	8.2
KP12X	Direct coupled	150/125	264)*	2950	6800	8.5	10.4[1]
Unistream	end suction		200)*	2950	6650	2.4	5.4
			264	2950)*	6800	8.5	10.4[1]
			264	2200)*	5700	4.7	5.8
KP10X	Direct coupled	125/100	264)*	2950	5500	7.6	9.8
Unistream	end suction		195)*	2950	4200	3.6	5.3
			264	2950)*	5500	7.6	9.8
			264	2200)*	5500	3.0	5.7
KM10E	Close coupled	125/100	264	2950	5500	7.6	9.8
Eurostream	end suction		195	2950	4200	3.6	5.3
KP08Y[4]	Direct	100/80	329)*	2950	3600	13.2	15.2[1]
Unistream	coupled end		270)*	2950	3500	7.0	10.0
	suction		255)*	2950	3450	5.6	8.8
			329	2950)*	3600	13.2	15.2[1]
			329	2400)*	3650	7.0	10.0
			329	2200)*	3650	5.0	8.3
KPO8V	Direct coupled	100/80	214	2920	2700	2.4	5.6

Continued

SPP Pumps (continued)

Model	Type	Suction Inlet Diameter/ Discharge Outlet Diameter (mm)	Impeller Diameter (mm)	Rated Speed (rev/min)	Rated Flow (L/min)(at NPSH req. = 5.38m)	Rated Head (bar)(at NPSH req. = 5.38m)	Closed Valve Head (bar)
Unistream	end suction		165	2920	2340	1.6	3.3
KMO8V	Close coupled	100/80	214	2920	2700	2.4	5.6
Eurostream	end suction		165	2920	2340	1.6	3.3
KPO6C	Direct coupled	80/65	174	2900	2590	2.3	4.3
Unistream	end suction		125	2900	1790	0.6	2.0
KMO6C	Close coupled	80/65	174	2900	2590	2.3	4.3
Eurostream	end suction		125	2900	1790	0.6	2.0
AVO3N	Direct coupled	100/80	174	2900	2540	1.3	3.8
Unistream	end suction		135	2900	1840	1.0	2.3
AVO3P	Close coupled	100/80	174	2900	2540	1.3	3.8
Eurostream	end suction		135	2900	1840	1.0	2.3
KPO5D	Direct coupled	65/50	214	2900	1380	3.3	6.3
Unistream	end suction		165	2900	1090	1.6	3.7
KMO5D	Close coupled	65/50	214	2900	1380	3.3	6.3
Eurostream	end suction		165	2900	1090	1.6	3.7
KP06D	Direct coupled	80/65	214)*	2950	2580	5.0	6.6
Unistream	end suction		165)*	2950	2250	2.3	3.8
			214	2950)*	2580	5.0	6.6
			214	2200)*	2550	1.7	3.6
KM06D	Close coupled	80/65	214	2950	2580	5.0	6.6
Eurostream	end suction		165	2950	2250	2.3	3.8
KP08D	Direct coupled	100/80	214)*	2950	3400	4.8	6.5
Unistream	end suction		165)*	2950	2820	2.4	3.8
			214	2950)*	3400	4.8	6.5
			214	2200)*	3300	1.8	3.7
KM08D	Close coupled	100/80	214	2950	3400	4.8	6.5
Eurostream	end suction		165	2950	2820	2.4	3.8

}* These pumps are approved with impeller diameters and, where applicable, speeds within the ranges shown, to provide flows and heads within those ranges.

[1.] Pump configurations which generate a closed valve head exceeding 10 bar are approved for use in accordance with TB13: 1990: 1 - High rise sprinkler systems.

[2.] Maximum working pressure: 21 bar

[3.] Maximum working pressure: 20 bar.

[4] Maximum working pressure : 18 bar.

The above pumps are fitted with "closed head" relief valve Pt. No. 25652 - 511 - 000 - 7, Pt. No. 25652-501-000-2.

The following multistage fire pump is LPCB approved for use in accordance with the *LPC Rules for automatic sprinkler installations,* Technical Bulletins:

TB12 : 1990 : 1 - Automatic Sprinkler Protection of Multi-storey Buildings, and

TB13 : 1990 : 1 - High-rise Sprinkler Systems.

Firms listed in this Section shall control and be responsible for the design, construction, testing and performance of all fire pump sets incorporating their LPCB approved fire pumps.

All fire pump sets incorporating LPCB approved fire pumps shall conform to the *LPC Rules for automatic sprinkler installations,* or Rules specified in the contract, and shall be tested in accordance with BS 5306 : Part 2, clause 17.4.13.11.

Note: Only LPCB approved fire pumps shall be used to comply with the *LPC Rules for automatic sprinkler installations.*

SPP Pumps
Sterling Fluid Systems Limited
Theale Cross, Reading, Berkshire RG3 7SP
(Also at Crucible Close, Coleford, Gloucester GL16 8PS)

Tel: +44 (0)1189 323123 • Fax: +44 (0)1189 323302

LPCB
LOSS PREVENTION
CERTIFICATION BOARD

Certificate No. 111b to LPS 1131/G

Model and Type	Suction Inlet Diameter/ Discharge Outlet Diameter (mm)	Delivery Configuration	Impeller Diameter (mm)	Rated Speed (rev/min)	Rated Flow (1/min) (at NPSH req. ≤5.38m)	Rated Head (bar) (at NSPH req. ≤5.38m)	Closed Valve Head (bar)
CD10K Multistream 2-5 stage multistage pump	125/100	(1) Single stage - closed stage or final outlet	219 202	2950 2950	2800 2800	4.8 3.3	7.0 5.5
		(2) single stage - intermediate outlet	219 202	2950 2950	2800 2800	3.7 1.4	7.5 5.9
		(3) Final outlet of five stage pump	219 202	2950 2950	2800 2800	24.0 16.5	35.0 27.5

Note: The pump performance is the summation of each stage, i.e. the performance at the final outlet of a four stage pump is 4 x (1).
The performance at the fourth stage intermediate outlet of a five stage pump is 3 x (1) +(2).
The performance at the final outlet of a five stage pump is 5 x (1), as detailed in (3).
The above pump is approved with impeller diameters within the range shown, to provide flows and heads within the ranges shown.
The above pump is fitted with 'closed head' relief valve part no. 25652-511-000-7.

5 SECTION 18:
PIPEWORK SUPPORTS

Pipework support requirements are specified in Section 22 of the *LPC Rules for automatic sprinkler installations.*

Components of pipe support systems listed below are deemed to have met the necessary performance requirements and are considered acceptable for use in installations conforming to the LPC Rules.

Debro Engineering and Presswork
Stourvale Trading Estate, Banners Lane, Halesowen B63 2AX

Tel: +44 (0)1384 636359 • Fax: +44 (0)1384 633746

Component : Filbow

Model	Pipe Size(mm)	Material	Finish
Debro Flat Filbow	20	Mild steel	Galvanised
Debro Flat Filbow	25	Mild steel	Galvanised
Debro Flat Filbow	32	Mild steel	Galvanised
Debro Flat Filbow	40	Mild steel	Galvanised
Debro Flat Filbow	50	Mild steel	Galvanised
Debro Flat Filbow	65	Mild steel	Galvanised
Debro Flat Filbow	80	Mild steel	Galvanised
Debro Flat Filbow	100	Mild steel	Galvanised
Debro Flat Filbow	150	Mild Steel	Galvanised
Debro Flat Filbow	200	Mild Steel	Galvanised

Grinnell Corporation
2870 Old Jackson Road, Henderson, Tennessee 38340, USA

Tel: +1 901 989 3551 • Fax: +1 901 989 4144
Sales Enquiries: Tel: +31 53 283434 • Fax: +31 53 283377
Also at 2785 Old Jackson Road, Henderson, Tennessee, USA

Component: Swivel band hanger

Model	Size (mm)	Material	Finish
70E	20	Mild steel	Galvanised
70E	25	Mild steel	Galvanised
70E	32	Mild steel	Galvanised
70E	40	Mild steel	Galvanised
70E	50	Mild steel	Galvanised
70E	65	Mild steel	Galvanised
70E	80	Mild steel	Galvanised
70E	100	Mild steel	Galvanised
70E	125	Mild steel	Galvanised
70E	150	Mild steel	Galvanised
70E	200	Mild steel	Galvanised

Grinnell Corporation
1411 Lancaster Avenue, Columbia, Pennsylvania 17512, USA

Tel: +1 717 684 4400 • Fax: +1 717 684 2131
Sales Enquiries: Tel: +31 53 428 3434 • Fax: +31 53 428 3377

Pipe Hanger Clamps
Figure 90 Universal C-Clamps M10 and M12 rod sizes[1,2,3]
Figure 90E Universal C-Clamps M16 rod sizes[1,2,3]
Figure 91 Universal C-Clamps M10 and M12 rod sizes[1,2,3]

[1] Rod drilled and rod threaded versions.
[2] Plain and galvanised finishes.
[3] A restraining strap shall be used where significant vibration is likely to be encountered.

Grinnell Manufacturing (UK) Ltd
Stockport Trading Estate, Yew Street, Stockport SK4 2JW

Tel: +44 (0)161 477 1886 • Fax: +44 (0)161 477 6729

Pipe Hanger Clamps
Standard Throat
SM10, M10 rod sizes[1,2,3]
SM12, M12 rod sizes[1,2,3]
SM16, M16 rod sizes[2,3,4]

Wide Throat
WM10, M10 Rod Sizes [1,2,3]
WM12, M12 Rod Sizes [1,2,3]

[1] Rod drilled and rod threaded versions
[2] Plain and galvanised finish
[3] A restraining strap shall be used where significant vibration is likely to be encountered
[4] Threaded versions only

Minimax GmbH
Industriestrasse 10/12, D-23840 Bad Oldesloe, PO Box 1260, Germany

Tel: +49 4531 8030 • Fax: +49 4531 803248
UK Sales Enquiries: Tel: +44 (0)181 832 2000 • Fax: +44 (0)181 832 2200

Component: Pipe Hanger

Model	Size (mm)	Material	Finish
SM	20	Mild Steel	Galvanised
SM	25	Mild Steel	Galvanised
SM	32	Mild Steel	Galvanised
SM	40	Mild Steel	Galvanised
SM	50	Mild Steel	Galvanised
SM	65	Mild Steel	Galvanised
SM	80	Mild Steel	Galvanised
SM	100	Mild Steel	Galvanised
SM	125	Mild Steel	Galvanised
SM	150	Mild Steel	Galvanised
SM	200	Mild Steel	Galvanised

SECTION 19:
CHECK VALVES

Check valves listed below are considered suitable for use under para 20.2 of the *LPC Rules for automatic sprinkler installations*. The products shall be installed and maintained in accordance with the manufacturer's instructions.

Grinnell Corporation

PO Box 128, Highway 70, Cleveland (Rowan County), North Carolina 27013, USA

Tel: +1 704 278 2221 • Fax: +1 704 278 9617
Sales Enquiries: Tel: +31 53 428 3434 • Fax: +31 53 428 3377

Model	Size (mm)	Body Material	Maximum Working Pressure (bar)	Attitude	Connections
Gruvlok Series 7800FP Figure GLV 7811 FP	50	Cast Iron	17	Vertical & horizontal	Grooved
	65	Cast Iron	17	Vertical & horizontal	Grooved
	80	Cast Iron	17	Vertical & horizontal	Grooved
	100	Cast Iron	17	Vertical & horizontal	Grooved
	150	Cast Iron	17	Vertical & horizontal	Grooved
	200	Cast Iron	17	Vertical & horizontal	Grooved
	250	Cast Iron	17	Vertical & horizontal	Grooved
	300	Cast Iron	17	Vertical & horizontal	Grooved

Stop valves listed below are considered suitable for use under para 20.1 of the *LPC Rules for automatic sprinkler installations.* The products shall be installed and maintained in accordance with the manufacturer's instructions.

Tyco Valves & Controls*

(A Tyco International Limited Company)
9700 West Gulf Bank Road, PO Box 40010, Houston, TX 77040, USA

Tel: +1 713 937 5342 • Fax: +1 713 937 5457
Sales Enquiries: Tel: +31 53 428 3434 • Fax: +31 53 428 3377

Gear operated stop valve with or without electrical monitor switches.

Model	Size (mm)	Body Material	Maximum Working Pressure (bar)	Connections
Gruvlok 7700	50	SG Cast Iron	12	Grooved
FP Butterfly	65	SG Cast Iron	12	Grooved
	80	SG Cast Iron	12	Grooved
	100	SG Cast Iron	12	Grooved
	125	SG Cast Iron	12	Grooved
	150	SG Cast Iron	12	Grooved
	200	SG Cast Iron	12	Grooved
	250	SG Cast Iron	12	Grooved
	300	SG Cast Iron	12	Grooved

* This company has not been Certificated by the LPCB to ISO 9000

INTRODUCTION

The 'scope of use' of plastic pipe shall be agreed between the plastics suppliers, purchaser/installer, authority having jurisdiction, and/or insurer in accordance with documented supplier 'Installation Instructions' but subject to the following criteria taking precedence.

1. Use of plastic pipe and fittings is subject to water authority agreement for the territory concerned.
2. LPCB Approved quick response sprinklers shall be used with exposed (i.e. fire exposure) plastic pipe and fittings.
3. Plastic pipe and fittings are suitable for use only with wet pipe systems.
4. Care should be exercised to ensure that joints are adequately cured, in accordance with the manufacturer's installation instructions, prior to pressurisation.
5. Plastic pipe and fittings shall not be installed outdoors.
6. Where plastic pipe and fittings are exposed (i.e. fire exposure), the system shall be installed close to a flat ceiling construction.
7. Sprinkler systems which employ plastic pipe and fittings shall be designed where possible to ensure no 'no flow' sections of pipework in the event of sprinkler operation.

Central CPVC Company

245 Swancott Road, Madison, Alabama 35756, USA

Tel: +1 256 464 5633 • Fax: +1 256 464 5635
UK Sales Enquiries: Tel: +44 (0)1427 615401 • Fax: +44 (0)1427 610433

CENTRAL SPRINKLER BLAZEMASTER® CHLORINATED POLYVINYLCHLORIDE (CPVC) SPRINKLER FITTINGS
Blazemaster ® is a registered trademark of BF Goodrich Limited

Product	Nominal diameter range	Maximum working pressure (bar)
Cap	¾",1",1¼",1½",2"	12
Straight Coupling	¾",1",1¼",1½",2"	12
Reducer Bush	1"x ¾",1¼"x¾",1¼"x1",1½"x¾",1½x1",1½x1¼"	12
90° Elbow	¾",1",1¼",1½",2"	12
45° Elbow	¾",1",1¼", 1½"	12
Tee	¾",1",1¼",1½",2"	12
Reducing Tee	1"x¾",1¼"x1¼x¾,1¼"x1¼"x1",1½"x1½x1"	12
Cross	¾",1",1¼",1½"	12
Slip sprinkler head adapters	¾"x½",1"x½"	12
Male/Female adaptor Coupling	1¼"x1"	12
Reducer Coupling	1"x¾"	12
Spigot sprinkler head Adaptor	¾"x½" & 1"x½"	12
Reducing bush	2"x1"	12
Reducing bush	2"x1¼"	12
Reducing bush	2"x1½"	12
Reducing Tee	1¼"x1"x1"	12
Reducing Tee	2"x2"x1"	12
Cross	2"	12
Sprinkler head 90° Elbow	1"x½"	12
Sprinkler head tee	1"x1"x½"	12

CENTRAL SPRINKLERS BLAZEMASTER® CHLORINATED POLYVINYLCHLORIDE (CPVC) SPRINKLER PIPE

Pipe	¾",1",1¼",1½", 2",2½"	12

Notes relating to Central pipes and fittings
1. The system of pipe and fittings must be installed in accordance with the Central Company "Blazemaster" Installation instructions, which includes LPCB Conditions of Use agreed with Central.
2. The maximum normal ambient temperatures of use shall not exceed 50°C.
3. The products shall only be installed by LPCB Certificated or Registered Installing companies (see Part 5, Section 1A and 1B above) or by firms outside the UK who can provide evidence of personnel training in the installation of the product. It is recommended that firms engaged in the installation of this product also be LPCB Certificated to ISO 9000.

Hershey Valve Co Limited*

No 356 Tzuli Road, Wuchi, Taichung, Taiwan

Tel: +886 4 630 0465 • Fax: +886 4 630 0467

Product	Nominal diameter range	Maximum working pressure (bar)
Tee	¾", 1", 1¼", 1½", 2"	12
Cross	¾", 1", 1¼", 1½", 2"	12
Reducing cross	1"x ¾"	12
90° elbow	¾", 1", 1¼", 1½", 2"	12
45° elbow	¾", 1", 1¼", 1½", 2"	12
Straight coupling	¾", 1", 1¼", 1½", 2"	12
Cap	¾", 1", 1¼", 1½", 2"	12
Female head adaptor	1"x ½"	12
Reducer bush	1¼"x ¼",1¼"x ¾", 1¼"x1",1½"x ½",	
	1½"x ¾",1½"x1",1½"x1¼",2"x1½"	12
Spigot sprinkler head adaptor	¾"x ½" and 1"x ½"	12
Male adaptor	½"x ½"	12
90° elbow male head adaptor	½"x ½"	12
90° elbow female head adaptor	½"x ½" and 1"x ½"	12
Sprinkler head tee	1"x1"x ½"	12
Pipe	¾",1",1¼",1½", 2"	12

* This company has not been certificated by the LPCB to ISO 9000.

Harvel Plastics Inc.

300 Kuebler Road, Easton, PA 18040, USA

Tel: +1 610 252 7355 • Fax: +1 610 253 4436

HARVEL BLAZEMASTER® 2000™ CHLORINATED POLYVINYLCHLORIDE (CPVC) SPRINKLER PIPE

Blazemaster ® is a registered trademark of BF Goodrich Chemical (UK) Limited

Pipe nominal diameter ¾", 1", 1¼", 1½", 2", 2½", 3"   Maximum working pressure 12 bar

Notes relating to Spears fittings and Harvel pipes

1   The system of pipe and fittings must be installed in accordance with Harvel Plastics Inc. **and** Spears Manufacturing Company, 'Blazemaster' Installation Instructions, which includes LPCB Conditions of Use agreed with Harvel and Spears.

2   The maximum normal ambient temperature of use shall not exceed 50˚C.

3   The products shall only be installed by LPCB Certificated or Registered installing companies (see Part 5, Section 1A and 1B above), or by firms outside the UK who can provide evidence of personnel training in the installation of the product. It is recommended that firms engaged in the installation of this product also be LPCB Certificated to ISO 9000.

Spears Manufacturing Co. Inc.
15853 Olden Street, Sylmar, CA 91342, USA
Tel: +1 818 364 1611 • Fax: +1 818 364 6945

SPEARS BLAZEMASTER® CHLORINATED POLYVINYLCHLORIDE (CPVC) FIRE SPRINKLER FITTINGS
Blazemaster ® is a registered trademark of BF Goodrich Chemical (UK) Limited

Product	Nominal diameter range	Maximum working pressure (bar)
Tee	¾",1",1¼",1½",2",2½",3"	12
Reducing Tee	¾"x¾"x1",1"x¾"x¾",1x¾"x1",	12
	1¼"x1"x¾",1¼"x1"x1",	
	1¼"x1"x1¼",1¼"x1¼"x¾",1¼"x1¼x1",	
	1¼"x1¼"x1½",1½"x1¼"x¾",	
	1½"x1¼"x1",1½"x1½"x¾",	
	1½"x1½"x1",1½"x1½"x1¼",	
	1½"x1½"x2",2"x2"x¾",2"x2"x1",2"x2"x1½",	
	2½"x2½"x1",2½"x2½"x1¼",	
	2½"x2½"x1½",3"x3"x2",3"x3"x2½"	
Sprinkler Head Tee	¾"x¾"x½",1"x½"x1",1"x¾"x½",1"x1"x½"	12
	1"x1"x1",1¼"x1x½",1¼"x1¼"x½",	
	1½"x1¼"x½",1½"x1½"x½",2"x1½"x½",	
	2"x2"x½"	
90° Elbow	¾",1",1¼",1½",2",2½",3",1"x¾"	12
Sprinkler Head 90° Elbow	¾"x½",1"x½",1"x¾",1¼"x½"	12
45° Elbow	¾",1",1¼",1½",2"2½",3"	12
Cross	¾",1",1¼",1½",2"2½",3"	12
Coupling	¾",1",1¼",1½",2"2½",3"	12
Cap	¾",1",1¼",1½",2"2½",3"	12
Flange	¾",1",1¼",1½",2"2½",3"	12
Blind Flange	¾",1",1¼",1½",2"2½",3"	12
Spigot Flange	¾",1",1¼",1½",2"2½",3"	12
Reducer coupling	1"x¾",1¼"x¾",1¼"x1",1½"x¾",1½"x1"	12
	1½"x1¼",2"x1",2"x1¼",2"x1½"	
Grooved Coupling Adapter	1¼",1½",2",2½",3"	12
Female Sprinkler Head Adapter	¾"x½",1"x½",1"x¾"	12
Reducer Bushing	1"x¾",1¼"x¾",1¼"x1",1½"x¾",1½"x1",	12
	1½"x1¼",2"x¾",2"x1",2"x1¼",2"x1½"	
	2½"x1¼",2½"x1½",2½"x2",3"x2",3"x2½"	
Spigot Sprinkler Head Adapter	¾"x½",1"x½"	12
Union	¾",1",1¼",1½",2",3"	12
Spigot Female Adapter	¾",1"	12
Adjustable Sprinkler Head Adapter (slip)	¾"x½",1"x½"	12
Adjustable Sprinkler Head Adapter (spigot)	¾"x½",1"x½"	12

Notes relating to Spears fittings and Harvel pipes

1   The system of pipe and fittings must be installed in accordance with Harvel Plastics Inc. **and** Spears Manufacturing Company, 'Blazemaster' Installation Instructions, which includes LPCB Conditions of Use agreed with Harvel and Spears.

2   The maximum normal ambient temperature of use shall not exceed 50˚C.

3   The products shall only be installed by LPCB Certificated or Registered installing companies (see Part 5, Section 1A and 1B above), or by firms outside the UK who can provide evidence of personnel training in the installation of the product. It is recommended that firms engaged in the installation of this product also be LPCB Certificated to ISO 9000.

Note:

'Metron' actuators listed below have been evaluated in relation to their general performance capability but not in relation to their suitability for specific applications.

Guidance:

1. Care shall be exercised to ensure that the monitoring current specified by the manufacturer is not exceeded.
2. The manufacturer's installation instructions and product safety instructions shall be followed.
3. These devices shall be replaced every ten years.

Nobel Enterprises

Ardeer Site, Stevenston, Ayrshire, Scotland KA20 3LN

Tel: +44 (0)1294 487000 • Fax: +44 (0)1294 487370

Protractor 'Metron' actuators

Metron Type	Stroke (mm)	Work Output J (minimum)	Resistance Range Ohms	Allfire Amps (50ms)	Nofire Amps (5 sec)
DR2001/C13	3.4	0.69	0.9 to 1.6	0.6	0.3
DR2003/C1	3.4	1.18	0.9 to 1.6	0.6	0.3
DR2003/C6	3.4	1.18	0.9 to 1.6	0.6	0.3
DR2003/C35	3.4	1.18	0.9 to 1.6	0.6	0.3
DR2004/C1	9.5	3.43	0.9 to 1.6	0.6	0.3
DR2005/C1	9.5	3.43	0.9 to 1.6	0.6	0.3
DR2005/C2	9.5	3.43	0.9 to 1.6	0.6	0.3
DR2005/C3	9.5	3.43	0.9 to 1.6	0.6	0.3
DR2005/C6	9.5	3.43	0.9 to 1.6	0.6	0.3
DR2005/C9	9.5	3.43	0.9 to 1.6	0.6	0.3
DR2005/C35	9.5	3.43	0.9 to 1.6	0.6	0.3
DR2006/C1	13.5	4.9	0.9 to 1.6	0.6	0.3
DR2006/C2	13.5	4.9	0.9 to 1.6	0.6	0.3
DR2014/C1	9.5	4.9	0.9 to 1.6	0.6	0.3
DR2053	3.4	0.8	7 to 12	0.16	0.05

Maximum monitoring current for all models is 0.01A.

Grinnell Corporation Limited

PO Box 128, Highway 70, Cleveland (Rowan County), North Carolina 27013, USA

Tel: +1 704 278 2221 • Fax: +1 704 278 9617
Sales Enquiries: Tel: +31 53 428 3434 • Fax: +31 53 428 3377

Water Motor Alarm   Model F630

Minimax GmbH

Industriestrasse 10/12, D-23840 Bad Oldesloe, Germany

Tel: +49 4531 8030 • Fax: +49 4531 803 248

Water Motor Alarm   Model AG2

The Reliable Automatic Sprinkler Co Inc

525 North MacQuesten Parkway, Mount Vernon, New York 10552-2600, USA

Tel: +1 914 668 3470 • Fax: +1 914 668 2936
UK Sales Enquiries: Tel: +44 (0)1372 728899 • Fax: +44 (0)1372 724461

Water Motor Alarm Model C

Star Sprinkler Inc.

PO Box 128, Highway 70, Cleveland (Rowan County), North Carolina 27013, USA

Tel: +1 414 570 5000 • Fax: +1 414 570 5010

Water Motor Alarm   Model S450

SECURITY SYSTEMS

INTRODUCTION

Intruder detection equipment

Intruder detection equipment included in Section 1 has been evaluated to the relevant standards and a product quality audit of its manufacture is made twice a year by the LPCB for compliance with ISO 9001 or ISO 9002.

All approved equipment will bear the LPCB Approval Mark together with the LPC reference number.

TEST STANDARDS

Requirements are defined in the following Loss Prevention Standards:

Ultrasonic doppler detectors	- LPS 1167
Microwave doppler detectors	- LPS 1168
Passive infrared detectors	- LPS 1169
Combined technology intruder alarms movement detectors	- LPS 1188
Control and indicating equipment for intruder and hold-up alarms	- LPS 1200

Safe storage units

Safe storage units included in Section 2 have been evaluated to LPS 1183. Certification is maintained by product audits and ISO 9001 or 2 surveillances undertaken by product experts.

All approved equipment will bear the LPCB Certification mark together with its reference number.

LPS1183 is based on EN1143.

The class and number of locks for each grade of safe is defined in LPS1183. The locks approved for use on each LPCB certificated safe is available on request from the manufacturer.

Automatic teller machines (ATM)

The ATMs listed in Section 3 have been tested to the safe storage unit standard LPS 1183. This is more onerous than the European standard in terms of partial access. The mounting test in the European standard is undertaken in place of the anchoring test in LPS 1183.

The same scheme conditions apply as for Safe storage units detailed above.

Security doors and shutters

Security doors and shutters included in Section 4 have been evaluated to LPS1175. Certification is maintained by product audits and ISO 9001 or 2 surveillance undertaken by product experts.

All approved equipment will bear the LPCB Approval mark together with its reference number.

The security ratings given in this list are based on the premise that all opening elements are in the "locked condition" as defined in LPS1175. No allowance has been made with regard to the stages of locking, e.g. latched but not locked.

To ensure that the products give the required security rating they must be installed in accordance with the manufacturer's instructions and be in the "locked condition". It is advisable to regularly check that all exposed fixings and locking mechanisms are in good condition and that they have not been tampered with.

Physical protection devices for personnel computers

Physical protection devices for personal computers included in Section 5 have been evaluated to LPS 1214. Certification is maintained by ISO 9002 surveillance product audits.

All approved equipment will bear the LPCB Certification mark together with its reference number.

To ensure that the products give the required security rating they must be installed in accordance with the manufacturer's instructions and be in the locked condition. It is advisable to regularly check that all exposed fixings and locking mechanisms are in good condition and that they have not been tampered with.

LPS 1214 Issue 1 and Issue 2 - Applications

LPS 1214 Issue 2 - Security Category 1 or LPS 1214 Issue 1 (Lock down)

Prevent unauthorised removal of personal computers (PCs), file servers, printers, facsimile machines and other high value equipment for example where access is important for use and the component value low.

LPS 1214 Issue 2 - Security Category 2 (Enclosure)

Prevent unauthorised removal of or access to personal computers (PCs), file servers, printers, facsimile machines and other high value equipment for example where the PC is both important and highly valued and/or the component value is high.

Database management for asset marking

Companies included in Section 6 have been assessed to LPS 1224. Certification is maintained by ISO 9001 or 2 surveillances undertaken in conjunction with the requirements of LPS 1224. The sope of approval covers management and access to the asset database.

6

SECTION 1:
INTRUDER DETECTION EQUIPMENT

Detectors are generally in accordance with appropriate Loss Prevention Standard, but see 'Notes'.

Aritech BV

Roermond, Delfstoffen - Weg 2, 6074 NK Melick, Netherlands

Tel: +31 4750 73535 • Fax: +31 4750 26456

UK Sales Enquiries: Aritech UK Limited

Tel: +44 (0)1708 381496 • Fax: +44 (0)1708 381371

Approval Schedule No. 052a to LPS1169: Issue 2

Approved Products	LPCB Ref No.
EV225 passive infrared detector [1,2]	052a/01
EV226 passive infrared detector [1,2]	052a/02
EV228 passive infrared detector [1,2]	052a/03
EV235 passive infrared detector [1,2]	052a/05
EV236 passive infrared detector [1,2]	052a/06
EV238 passive infrared detector [1,2]	052a/07
EV635 passive infrared detector [a,1,2]	052a/08
EV636 passive infrared detector [a,1,2]	052a/09
EV645 passive infrared detector [a,1,2]	052a/10
EV646 passive infrared detector [a,1,2]	052a/11
EV227 passive infrared detector [1,2]	052a/12
EV425 plus volumetric passive infrared detector	052a/13
EV125 plus curtain passive infrared detector	052a/14
EV 455 plus Curtain passive infrared detector	052a/15

Notes: [a] - Not approved in '4D OFF' mode.
[1] - do not provide adequate protection against ingress of insects as per clause 6.4.1 of LPS 1169.
[2] - do not provide protection to IP41 as per clause 6.4.2 of LPS 1169.

Cerberus AG

Volketswil Factory, CH-8603 Schwerzenbach, Switzerland

Also Alte Landstrasse 411, CH-8708 Männedorf, Switzerland

Tel: +41 1 922 6111 • Fax: +41 1 922 6450

UK Sales:

Alarmcom Limited Tel: +44 (0)1245 478585 • Fax: +44 (0)1245 478530

Approval Schedule No. 127b to LPS1169: Issue 2

Approved Products	LPCB Ref No.	LPC Ref. No.
IR210C volumetric passive infra-red detector	127b/01	78/1
IR212C curtain passive infra-red detector	127b/02	78/2
IRI50B passive infared detector[1,2]	127b/03	78/3
Using following:		
volumetric mirror - supplied with product		
corridor mirror - IRS152		
mounting bracket - IRUM1		
IR 310 C volumetric passive infra-red detector[1,2]	127b/04	78/4
IR 312 C curtain passive infra-red detector[1,2]	127b/05	78/5
IR220B volumetric passive infrared detector	127b/06	
IR222B curtain passive infrared detector	127b/07	
IR220C volumetric passive infrared detector	127b/08	
IR222C curtain passive infrared detector	127b/09	
IR160B passive infrared detector[1,3]	127b/10	
Using following:		
volumetric mirror - supplied with product		
corridor mirror - IRS162		
mounting bracket - IRUM1		
IR 160B passive infrared detector[1,3]	127b/10	
Using the following:		
volumetric mirror - supplied with product		
corridor mirror - IRS 162		
mounting bracket - IRUM1		

Notes: [1] - do not provide adequate protection against ingress of insects as per clause 6.4.1 of LPS 1169.
[2] - do not provide protection to IP41 as per clause 6.4.2 of LPS 1169.
[3] - approved in 'high evaluation mode' only.

Detectors are generally in accordance with appropriate Loss Prevention Standard, but see '*Notes*'.

Guardall Limited

Lochend Industrial Estate, Newbridge, Edinburgh, Midlothian EH28 8PL, Scotland

Tel: +44 (0)131 333 2900 • Fax: +44 (0)131 333 4919

Approval Schedule No. 130a to LPS1169: Issue 2

Approved Products	LPC Ref No.
Apollo Elite/1 passive infrared detector [1,2] using the following lenses; volumetric - AP12/3, supplied with product curtain - AP10/3 long range - AP20/3	130a/01
Apollo Solo/1 passive infrared detector [1,2] using the following lenses; volumetric - AP12/3, supplied with product curtain - AP10/3 long range - AP20/3	130a/02

Notes: [1] - *do not provide adequate protection against ingress of insects as per clause 6.4.1 of LPS 1169.*
[2] - *do not provide protection to IP41 as per clause 6.4.2 of LPS 1169.*

Pyronix Limited

Unit 2, Braithwell Way, Hellaby Industrial Estate, Hellaby, Rotherham, South Yorkshire S66 8QY

Tel: +44 (0)1709 700100 • Fax: +44 (0)1709 701042

Approval Schedule No.160a to LPS 1169: Issue 2

Approved Products	LPCB Ref No.
Magnum LX passive infrared detector The following lens options are approved: lens 1 - standard volumetric lens 2 - high density volumetric lens 4 - long range	160a/01

Scantronic Limited

Perivale Industrial Park, Greenford, Middlesex UB6 7RJ

Tel: +44 (0)181 991 1133 • Fax: +44 (0)181 997 4448

Approval Schedule No. 240a to LPS 1169: Issue 2

Approved Products	LPCB Ref No.
1062 Passive infrared detector[1,2] Using the following lenses: - wide angle - corridor	240a/01

Notes: [1] - *do not provide adequate protection against ingress of insects as per clause 6.4.1 of LPS 1169.*
[2] - *do not provide protection to IP41 as per clause 6.4.2 of LPS 1169.*

Chubb Lips Nederland B.V.

Merwedestraat 48, 3313 CS, Postbus 59, 3300 AB Dordrecht, The Netherlands

Tel: +31 78 639 4041• Fax: +31 78 639 4305

Certificate No. 194a to LPS 1183 : Issue 4

LPCB Ref. No.	Trade Name	Size Reference	Security Grade
194a/02	Europlanet	150, 250, 350, 450 550, 650, 850, 950	II
194a/03	Europlanet	150, 250, 350, 450 550, 650, 850, 950	III
194a/04	Europlanet	100, 200, 300, 400 500, 700, 800, 801 802	IV

P T Chubb Lips Indonesia

Industrial Town MM2100, JL. Bali Blok T-6-1, Cibitung, Bekasi 17520, Indonesia

Tel: +62 21 898 1355 • Fax: +62 21 898 1371

Certificate No. 414a to LPS 1183 : Issue 4

LPCB Ref. No.	Trade Name	Size Reference	Security Grade
414a/06	Europa	1 to 5	0
414a/01	Europa	1 to 5	I
414a/02	Europa	1 to 7	II
414a/03	Europa	2 to 7	III
414a/04	Europa	3 to 7	IV
414a/05	Europa	3 to 7	V

Chubb Safe Equipment Company Limited

P. O. Box 61, Wednesfield Road, Wolverhampton, West Midlands WV10 0EW

Tel: +44 (0)1902 455111 • Fax: +44 (0)1902 450949

Certificate No. 114a to LPS 1183: Issue 3
Certificate No. 114b to LPS 1183: Issue 4

LPCB Ref. No.	Trade Name	Sizes References	Test Standard	Security Grade
114b/10	Eurosafe 0	0, 5, 10, 20 & 30	LPS 1183 : Issue 4	0
114b/11	Eurosafe I (Mk 2)	0, 5, 10, 20 & 30	LPS 1183 : Issue 4	I
114a/01	Eurosafe 1	10, 20, 30 and 50	LPS 1183 : Issue 3	I
114a/02	Eurosafe 2	10, 20, 30, 40 and 50	LPS 1183 : Issue 3	II
114a/03	Eurosafe 3	20, 30, 40 and 50	LPS 1183 : Issue 3	III
114a/04	Eurosafe 4	20, 30, 40, 50, 60 and 70	LPS 1183 : Issue 3	IV
114a/05	Eurosafe 5	20, 30, 40, 50, 60 and 70	LPS 1183 : Issue 3	V
114b/05	Eurosafe 5 EX	20, 30, 40, 50, 60 and 70	LPS 1183: Issue 4	V EX
114b/06	Eurosafe 6 EX	20, 30, 40, 50, 60 and 70	LPS 1183: Issue 4	VI EX

Note: The Eurosafe range LPCB Ref. 114a is also marketed under the Castelle name, Eurosafe 1 to 5 being equivalent to Castelle C+ to G+ respectively.

Churchill Safe & Security Products Limited

Brymbo Road Industrial Estate, Holditch, Newcastle-under-Lyme, Staffordshire ST5 9HZ.

Tel: +44 (0)1782 717400 • Fax: +44 (0)1782 711672

Certificate No. 152a to LPS 1183: Issue 4

LPCB Ref. No.	Trade Name	Size References	Security Grade
152a/01	Euro Domestic	ED2, ED3	0 *
152a/02	Euro Vector	EV2, EV3	I *
152a/03	Euro Extra	EE12, EE13	II *
152a/04	Euro Banker	EB22, EB23	III **
152a/05	Euro Olympic	EO2, EO3	0 *
152a/06	Euro Olympic	EO12, EO13	I *
152a/07	Euro Olympic	EO22, EO23	II *

* These safes can be surface mounted in addition to under floor.
** The manufacturer recommends that this safe is installed under floor only.

Dudley Safes Limited

Unit 17 Deepdale Works, Deepdale Lane, Upper Gornal, Dudley, West Midlands DY3 2AF

Tel: +44 (0)1384 239991 • Fax: +44 (0)1384 455129

Certificate No. 304a to LPS 1183 : Issue 4

LPCB Ref. No.	Trade Name	Size References	Security Grade
304a/01	Europa	2 to 7	0
304a/02	Europa	2 to 7	I
304a/03	Europa	2 to 7	II

Melsmetall GMBH

Spangenberger Strasse 61, D-34212 Melsungen, Germany

Tel: +49 5661 7350 • Fax: +49 5661 73544

Certificate No. 376a to LPS 1183 : Issue 4

LPCB Ref. No.	Trade Name	Product Type	Size References	Security Grade
376a/01	Safeguard	Underfloor Safe	1	0
			2	0
			3	0
376a/02	Protector Minor	Underfloor Safe	1	0
			2	0
			4	0
376a/03	Protector	Underfloor Safe	1	I
			2	I
			4	I
376a/04	Protector ABP	Underfloor Safe	1	II
			2	II
			4	II

6 SECTION 2:
SAFE STORAGE UNITS

Rosengrens Europe B.V.
Mercuriusstraat 60, 7006 RM Doetinchem, The Netherlands

Tel: +314 371600 • Fax: +31 4 327560

Certificate No. 181a to LPS 1183: Issue 3
Certificate No. 181b to LPS 1183: Issue 4

LPCB Ref. No.	Trade Name	Sizes (cm) Internal dimensions	Test Standard	Security Grade	Fire Resistance (min) BS 476:Part 20:1987
181b/10	Eurocitizen /protect	34 40 32, 51 40 32 70 40 32, 90 40 32 34 50 40, 51 50 40 70 50 40, 90 50 40 120 50 40, 160 50 40	LPS 1183 : Issue 4	0	60
181b/11	Eurocitizen /protect (LW)	34 40 32, 51 40 32 70 40 32, 90 40 32 34 50 40, 51 50 40 70 50 40, 90 50 40, 120 50 40, 160 50 40	LPS 1183 : Issue 4	0	60
181a/01	the European	50 40 35, 70 40 35, 90 40 35, 50 50 40, 70 50 40, 90 50 40, 120 50 40, 160 50 40, 50 50 50, 70 50 50, 90 50 50, 120 50 50, 160 50 50	LPS 1183 : Issue 3	I	60
181b/12	Eurocitizen /protect (LW)	34 40 32, 51 40 32 70 40 32, 90 40 32 34 50 40, 51 50 40 70 50 40, 90 50 40 120 50 40, 160 50 40	LPS 1183 : Issue 4	I	60
181b/13	Eurocitizen /protect	34 40 32, 51 40 32 70 40 32, 90 40 32 34 50 40, 51 50 40 70 50 40, 90 50 40 120 50 40, 160 50 40	LPS 1183 : Issue 4	I	60
181a/02	the European	50 40 35, 70 40 35, 90 40 35, 50 50 40, 70 50 40, 90 50 40, 120 50 40, 160 50 40, 50 50 50, 70 50 50, 90 50 50, 120 50 50, 160 50 50	LPS 1183 : Issue 3	II	60
181a/03	the European	50 50 40, 90 50 40, 120 50 40, 160 50 40, 50 50 50, 90 50 50, 120 50 50, 160 50 50	LPS 1183 : Issue 3	III	60
181a/04	the European	50 50 40, 90 50 40, 120 50 40, 160 50 40, 50 50 50, 90 50 50, 120 50 50, 160 50 50, 170 66 43, 170 97 43	LPS 1183 : Issue 3	IV	60
181b/05	the European	90 50 40, 120 50 40 160 50 40, 170 66 43 170 97 43	LPS 1183 : Issue 4	V	60
181a/06	the European	90 50 40, 120 50 40, 160 50 40, 90 50 50, 120 50 50, 160 50 50	LPS 1183 : Issue 3	VI	60

Rosengrens Produktions AB

Box 450, S-792 27 Mora, Sweden

Tel: +46 250 567800 • Fax: +46 250 567890

Certificate No. 267a to LPS 1183: Issue 4

LPCB Ref. No.	Trade Name	Sizes (cm) Internal dimensions	Security Grade	Fire Resistance (min) BS 476:Part 20:1987
267a/01	Eurocitizen/protect	34 40 32, 51 40 32, 70 40 32, 90 40 32, 34 50 40, 51 50 40, 70 50 40, 90 50 40, 120 50 40, 160 50 40	0	60 min
267a/02	Eurocitizen/protect(LW)	34 40 32, 51 40 32, 70 40 32, 90 40 32, 34 50 40, 51 50 40, 70 50 40, 90 50 40, 120 50 40, 160 50 40	0	60 min
267a/03	Eurocitizen/protect(LW)	34 40 32, 51 40 32, 70 40 32, 90 40 32, 34 50 40, 51 50 40, 70 50 40, 90 50 40, 120 50 40, 160 50 40	I	60 min
267a/04	Eurocitizen/protect	34 40 32, 51 40 32, 70 40 32, 90 40 32, 34 50 40, 51 50 40, 70 50 40, 90 50 40, 120 50 40, 160 50 40	I	60 min

SMP Security Limited

Halesfield 24, Telford, Shropshire TF7 4NZ

Tel: +44 (0)1952 585673 • Fax: +44 (0)1952 582816

Certificate No. 115a to LPS 1183: Issue 3
Certificate No. 115b to LPS 1183: Issue 4

LPCB Ref. No.	Trade Name	Size References	Test Standard	Security Grade	Fire Resistance BS 476: Part 20: 1987
115b/10	Community	1414 (39) ED (58) 1814 (50) ED (74) 2316 (103) ED (138) 2916 (130) ED (174) 3520 (230) ED (297) 5020 (328) ED (424)	LPS 1183 : Issue 4	0	60
115b/11	Community Mk2	1414 (39) ED (58) 1814 (50) ED (74) 2316 (103) ED (138) 2916 (130) ED (174) 3520 (230) ED (297) 5020 (328) ED (424)	LPS 1183 : Issue 4	0	60
115a/01	Community	1414 (39) ED (58) 1814 (50) ED (74) 2316 (103) ED (138) 2916 (130) ED (174) 3520 (230) ED (297) 5020 (328) ED (424) 6020 (393) ED (509) Mini-vault (702) ED (910)	LPS 1183 : Issue 3	I	60

Continued

6 SECTION 2:
SAFE STORAGE UNITS

SMP Security Limited (continued)

LPCB Ref. No.	Trade Name	Size References	Test Standard	Security Grade	Fire Resistance BS 476: Part 20: 1987
115a/02	Community	1814 (50) ED (74) 2316 (103) ED (138) 2916 (130) ED (174) 3520 (230) ED (297) 5020 (328) ED (424) 6020 (393) ED (509) Mini-vault (702) ED (910)	LPS 1183 : Issue 3	II	60
115a/03	Community	1814 (50) ED (74) 2316 (103) ED (138) 2916 (130) ED (174) 3520 (230) ED (297) 5020 (328) ED (424) 6020 (393) ED (509) Mini-vault (702) ED (910)	LPS 1183 : Issue 3	III	60
115a/04	Community	60 50 40 (120) ED (165) 90 50 40 (180) ED (247) 120 50 40 (240) ED (330) 150 50 40 (300) ED (412)	LPS 1183 : Issue 3	IV	—
115b/05	Community	60 50 40 (120) ED (165) 90 50 40 (180) ED (247) 120 50 40 (240) ED (330) 150 50 40 (300) ED (412)	LPS 1183 : Issue 4	V	—

Note:
(1) ED denotes available with up to 150mm extra depth.
(2) Figures in brackets denote the volumes in litres. The first figure is the standard volume, the second figure is the volume of safe with maximum 150mm extra depth.

Rosengrens Europe BV

Mercuriusstraat 60, 7006 RM Doetinchem, The Netherlands

Tel: +31 4 371600 • Fax: +31 4 327560

Certificate No. 181c to LPS 1183 : Issue 4 and amendment A1: 1997

LPCB Ref. No.	Trade Name	Size Reference	Security Grade
181c/01	Seceuro	5684 / 5884	A3
181c/02	Seceuro	5685 / 5885	A3
181c/03	Seceuro	TCD9310	A3

Alsecure Limited

Heath Hill, Dawley, Telford, Shropshire TF4 2RJ

Tel: +44 (0)1952 402020 • Fax: +44 (0)1952 402030

Certificate No. 219a to LPS 1175 Issue 3

LPCB Ref. No.	Trade Name	Type	Minimum Size (m)		Maximum Size (m)		Security Rating
			Height	Width	Height	Width	
219a/01	Ramstop	Roller shutter	2	1	3	3	4

Amber Doors Limited

Mason Way, Platts Common Industrial Estate, Hoyland, Barnsley, South Yorks S74 9TG

Tel: +44 (0)1226 351135 • Fax: +44 (0)1226 350176

Certificate No. 001b to LPS 1175 Issue 3

LPCB Ref.No.	Trade Name	Type	Size (m)		Security Rating
			Height	Width	
001b/01	Protector Series 4000 Security Shutter	Rolling Shutter	3 (Min) 5 (Max)	3 (Min) 7 (Max)	4

Notes: 1) This product must be face fixed to inside of premises.
2) A centre ground lock must be provided on shutters of greater than 5 metres width.

Ascot Industrial Doors Limited

Britannia Way Industrial Park, Union Road, Bolton BL2 2HE

Tel: +44 (0)990 556644 • Fax: +44 (0)1204 545800

Certificate No. 131b to LPS 1175 : Issue 3

LPCB Ref. No.	Trade Name	Type	Minimum Size (m)		Maximum Size (m)		Security Rating
			Height	Width	Height	Width	
131b/01	A320	Rolling Shutter	3	9	2	2.5	4

Bradbury Security Grilles

Dunlop Way, Queensway Enterprise Estate, Scunthorpe, DN16 3RN

Tel: +44 (0)1724 271999 • Fax: +44 (0)1724 271888

Certificate No. 312a to LPS 1175 Issue 3

LPCB Ref. No.	Trade Name	Type	Minimum Size (m)		Maximum Size (m)		Security Rating
			Height	Width	Height	Width	
312a/01	Crossguard C21	Security grille			3	4	1
312a/02		Steel Roller Shutter[1]	1	1	3	3	2

Note: [1] The certification covers this product when reveal or internal face fixed.

Chiltern Industrial Doors Limited
Unit 9, Commerce Way, Leighton Buzzard, Bedfordshire LU7 8RW

Tel: +44 (0)1525 383537 • Fax: +44 (0)1525 382314

Certificate No. 246b to LPS 1175 Issue 3

LPCB Ref. No.	Trade Name	Type	Sizes			Security Rating
			Height (m) (max)	Width (m) (min)	Width (m) (max)	
246b/01	Chiltern Security shutter	Roller shutter	2.5	1.5	2.5	2

N T Martin Roberts Limited
Millen Road, Sittingbourne, Kent ME10 2AA

Tel: +44 (0)1795 476161 • Fax: +44 (0)1795 422463

Certificate No. 322a to LPS 1175 Issue 3

LPCB Ref. No.	Trade Name	Type	Maximum Size		Security Rating Locked Condition
			Height (m)	Width (m)	
322a/01	3SD Series Guardian OMNI Secure Doorset	Single leaf hinged security door with or without emergency exit facilities	2.095	1.002	3

Security Window Shutters Ltd
Lansil Way, Lansil Industrial Estate, Lancaster LA1 3QY

Tel: +44 (0)1524 33986 • Fax: +44 (0)1524 844057

Certificate No. 390a to LPS 1175 Issue 3

LPCB Ref. No.	Trade Name	Type	Size				Security Rating
			Height (m) (max)	Width (m) (max)	Height (m) (min)	Width (m) (min)	
390a/01	SeceuroGuard 1001	Collapsible Single Sash Security Gate	3	3	1	0.5	1
		Collapsible Double Sash Security Gate	3	6	1	1	1

Sunray Engineering Limited

Kingsnorth Industrial Estate, Wotton Road, Ashford, Kent TN23 6LL

Tel: +44 (0)1233 639039 • Fax: +44 (0)1233 625137

Certificate No. 234a to LPS 1175 Issue 3

LPCB Ref. No.	Trade Name	Type	Size Height (m)	Width (m)	Security Rating
234a/01	ExcluDoor 4 with panic bar	Hinged security door outward opening	2.25 (max)	1.05 (max)	4
234a/02	ExcluDoor Saver Mk II[1]	Hinged security door outward opening	2.25 (max)	1.1 (max)	3
234a/03	ExcluDoor 4 2 Lock version	Hinged security door outward opening	1.8 (min) 2.25 (max)	0.75 (min) 1.05 (max)	4

Notes: [1]The above security rating would not be achieved if the key is left in the lock on the inside of the door or if the door is placed in the emergency exit mode.

AP Computer Security
Farndon Business Centre, Farndon Road, Market Harborough, Leicester LE16 9NP

Tel: +44 (0)1858 469500 • Fax: +44 (0)1858 468660

Certificate No. 302a to LPS 1214 : Issue 2

| LPCB Ref. | Model | Plate Size (mm) | | | Security category |
		Height	Width	Depth	
302a/01	1416N Lockdown Plate[1]		400	350	I
302a/02	Universal Enclosure System (Slimline) [2]	75-135	300-540	330-420	I
	Universal Enclosure System (Standard) [2]	135-195	300-540	330-420	I
302a/03	1418 Lockdown Plate [1]		360	338	I

Notes: 1. The dimension refers to the size of plate
2. The dimension refers to the size of enclosure
3. The approval applies to the products when installed strictly in accordance with the instructions provided with the product.

Argonaut Manufacturing
13 Nutwood Way, Westwood Business Park, Totton, Southampton SO40 3WW

Tel: +44 (0)1703 860990 • Fax: +44 (0)1703 870652

Certificate No. 326a to LPS 1214 : Issue 2

| LPCB Ref. No. | Model | Size (mm) | | | Security category |
		Height	Width	Depth	
326a/01	Strongbox				
	DTS	155	450	450	I & II
	DTM	155	500	500	I & II
	DTL	215	500	550	I & II
	DTXL	270	525	625	I & II
	MTL	500	215	550	I & II
	MTXL	525	270	625	I & II
326a/02	Tower Entrapment				
	MTM-1	495	305	550	I
	MTL-1	685	320	600	I
	MTXL-1	945	320	700	I

Notes : 1. The approval also covers any size within the maxima and minima quoted above.
2. The approval applies to these products when installed strictly in accordance with the instructions provided with the product.
3. The substrate and fixings used for floor mounting may affect the performance of the product and are therefore outside the scope of the approval.

Computer Security Systems

Unit 52, Offerton Industrial Estate, Hempshaw Lane, Offerton, Stockport SK2 5TJ

Tel: +44 (0)161 477 5807 • Fax: +44 (0)161 477 6940 - Head Office

Certificate No. 329a to LPS 1214 : Issue 2

LPCB Ref. No.	Model	Encased Computer Size (mm)				Security category
		Max Height Tower	Max Width Desktop	Min Height Desktop	Min Width Tower	
329a/01	Anchorpad Challenger II	600	600	75	160	I

Notes: 1. The approval applies to the product when installed strictly in accordance with the instructions provided with the product.
2. The substrate and fixings for attachment of this product to surfaces other than a desk may affect the performance and are therefore outside the scope of the approval.

Dalen Limited

Valepits Road, Garretts Green Industrial Estate, Birmingham B33 0TD

Tel: +44 (0)121 783 3838 • Fax: +44 (0)121 784 6348

Certificate No. 276a to LPS 1214 : Issue 2

LPCB Ref. No.	Model	Computer Size (mm)						Test Standard
		Maximum			Minimum			
		Height	Width	Depth	Height	Width	Depth	
276a/01	Top Tec T-T450/1	380	190	440	330	n/a	n/a	LPS 1214 : Issue 2 : Category 1
276a/02	Top Tec T-T450/2	380	190	440	430	n/a	n/a	LPS 1214 : Issue 2 : Category 1
276a/03	Top Tec T-T460/2	180	515	700	100	300	355	LPS 1214 : Issue 2 : Category 1

Notes: 1. The approval applies to these products when installed strictly in accordance with the instructions provided with the product.
2. The substrate and fixings used for wall and floor mounting may affect the performance of the product and are therefore outside the scope of the approval.

PC Protect Limited

Data House, 31 Craneswater, Harlington, Hayes, Middlesex UB3 5HW

Tel: +44 (0)181 384 2222 • Fax: +44 (0)181 384 2223

Certificate No. 294a to LPS 1214: Issue 2

LPCB Ref. No.	Model	Model	Enclosed computer size (inches)						Security category
			Maximum			Minimum			
			Height	Width	Depth	Height	Width	Depth	
294a/01	Fortress	MT820	22	8.5	20				I
		MT8100	22	9.4	17.4				
		MT921	22	9.2	21.0				
		MT922	22	9.5	22				
		MT924	22	9.5	24				
294a/02	Fortress	LDD1316	8	14.5	16	3	12.5	16	I
		LDD1416	8	15.5	16	3	13.5	16	
		LDD1516	8	16.5	16	3	14.5	16	
		LDD1716	8	19.5	16	3	16.5	16	

Notes: 1. The approval applies to these products when installed strictly in accordance with the appropriate instructions.
2. The substrate and fixings used for wall and floor mounting may affect the performance of the product and are therefore outside the scope of the approval.

Pro-Techt Computer Security Limited

Pro-Techt House, Raven Industrial Estate, Garnant, Ammanford,
Carmarthenshire SA18 1NS

Tel: +44 (0)1269 825884 • Fax: +44 (0)1269 824198

Certificate No. 354a to LPS 1214 : Issue 2

LPCB Ref. No.	Model	Enclosed computer size (mm)						Security category
		Maximum			Minimum			
		Height	Width	Depth	Height	Width	Depth	
354a/01	Gard-IT	750	370	750	475	320	495	I & II
354a/02	Protechtor	215	508	590	115	389	395	I
354a/03	Protechtor Minitower	515	270	540	385	190	435	I

Notes: 1. The approval applies to these products when installed strictly in accordance with the instructions provided with the product.
2. The approval covers the units only when fixed to a suitable work surface or desk side. Fixings to any other surface are outside the scope of the approval.

The Mighty Box Co Limited

Middlewatch, Swavesey, Cambridgeshire CB4 5RN

Tel: +44 (0)1954 206602 • Fax: +44 (0)1954 206601

Certificate No. 311a to LPS 1214 : Issue 2

LPCB Ref. No.	Trade Name	Product Code	Size (mm)			Security category
			Height	Width	Depth	
311a/01	Sentry	1D05	210	420	520	I
		1D10	250	542	580	I
		1T10	482	290	580	I
		1T20	782	290	580	I
		1D15	260	562	580	I
		1T15	660	330	580	I
		1S30	782	380	795	I
		1S40	842	500	840	I
311a/02	Fortress	2D05	210	420	520	I & II
		2D10	250	542	580	I & II
		2T10	482	290	580	I & II
		2T20	782	290	580	I & II
		2D15	260	562	580	I & II
		2T15	660	330	580	I & II
		2S30	782	380	795	I & II
		2S40	842	500	840	I & II

Notes: 1. The approval applies to these products when installed strictly in accordance with the instructions provided with the product.
2. The substrate and fixings used for wall and floor mounting may affect the performance of the product and are therefore outside the scope of the approval.

SECTION 6:

DATABASE MANAGEMENT FOR ASSET MARKING

Kodit Database Limited

Chapel House, 84 Chapel Lane, Wilmslow, Cheshire SK9 5JH

Tel: +44 (0)1625 529283 • Fax: +44 (0)1625 539833

Approval certificate No: 379a to LPS 1224

FIXED FIRE FIGHTING SYSTEMS

7 SECTION 1:
LPS 1204 CERTIFICATED FIXED SYSTEM FIRMS

INTRODUCTION

LPS 1204: Requirements for firms engaged in the design, installation and commissioning of fixed fire fighting systems. Sprinkler and water spray systems are not included.

The schemes identified below are now operational. In addition it is planned to shortly introduce a scheme *LPS 1205 Requirements for firms engaged in the servicing of fire fighting systems.* A combination of LPS 1204 and LPS 1205 will ensure that systems are designed, installed, commissioned and serviced in the proper manner to safeguard life and property.

LPS 1204 Certificated Firms (Section 1A)

Firms listed in Section 1A are authorised by the LPCB to Certificate fixed systems to the LPCB recognised specifications identified in LPS 1204.

Each firm is also LPCB certificated to ISO 9001 or ISO 9002 - *'Quality Systems'*, as appropriate.

Additionally, firms which are Certificated Supervising Firms are identified. These firms are authorised by the LPCB to supervise Registered Firms.

LPS 1204 Registered Firms (Section 1B)

Firms to be listed in this forthcoming Section 1B will be Registered Firms, recognised by the LPCB to design and install under the supervision of an LPCB Certificated Firm.

It is anticipated that some firms will achieve Registration during 1997 and will therefore appear in a future amendment to this publication.

Each firm will also be LPCB certificated to ISO 9001 or ISO 9002.

Certification of Fixed Systems and LPCB Certificates of Conformity

It is a requirement of LPS 1204 that Certificated Firms shall issue an LPCB Certificate of Conformity in respect of each completed installation, extension and alteration.

Additionally, a Certificated Supervising Firm shall issue an LPCB Certificate of Conformity in respect of each completed installation, extension and alteration which is designed and installed by a Registered Firm.

The issue of an LPCB Certificate of Conformity:

(a) certifies that the fixed system, extension and alteration is designed, installed and commissioned in accordance with the LPCB recognised standards identified in LPS 1204 and

(b) ensures that the certificated system, extension and alteration is recorded by the LPCB.

Note to Insurers, Specifiers and Clients

The purpose of the LPS 1204 scheme is to provide an assurance that fixed fire systems, extensions and alterations which are designed and installed by LPS 1204 Certificated and Registered firms conform to LPCB recognised standards.

A key element of the LPS 1204 scheme is the **LPCB Certificate of Conformity**, which is issued upon completion of the works. It is most important that these are requested at the time of contract, to provide the assurance that fixed systems, extensions and alterations are designed and installed in accordance with LPCB recognised standards.

Additional Offices

Where a firm has additional offices, each office is required to meet the requirements of LPS 1204 to be eligible for certification, and is separately listed.

LPCB Inspection of Fixed Systems

During the LPCB six-monthly follow-up inspection of Certificated and Registered Firms completed fixed systems are selected and inspected for conformity with the specified recognised standards. This is in addition to ISO 9001/9002 surveillances.

Recommendation to Users/Specifiers/Insurers/Interested Parties

It is recommended that the responsibility of the supplier for provision of the system in total, (including the integration of the following facets), be established at the contract stage:

• Risk assessment.
• Extinguishing design concentration.
• Enclosure structural/leakage integrity.
• Discharge/Post discharge venting.
• Electrical system integration.

7 SECTION 1:
LPS 1204 CERTIFICATED FIXED SYSTEM FIRMS

Note : Sprinkler and water spray systems are not included.

ADT Fire & Security
Unit 301 Century Building, Tower Street, Liverpool L3 4BL

Tel: +44 (0)151 707 1223 • Fax: +44 (0)151 709 1822

Certificate No. CFSI-006

SCOPE: Certificated to design, install and commission FM200 and Inergen systems and issue LPCB Certificates of Conformity for these systems

Ginge Kerr
526 Fleet Lane, St Helens, Merseyside WA9 2NB

Tel: +44 (0)1744 28671 • Fax: +44(0)1744 20174

Certificate No. CFSI-007

SCOPE: Certificated to design, install and commission Argonite systems and issue LPCB Certificates of Conformity for these systems.

How Fire Limited
Hillcrest Business Park, Cinderbank, Dudley, West Midlands DY2 9AP

Tel: +44 (0)1384 458993 • Fax: +44 (0)1384 458981

Certificate No. CFSI-005

SCOPE: Certificated to design, install and commission CO_2 systems and issue LPCB Certificates of Conformity for these systems.

Preussag Fire Protection Limited
Field Way, Greenford, Middlesex UB6 8UZ

Tel: +44 (0)181 832 2000 • Fax: +44 (0)181 832 2200

Certificate No. CFSI-004

SCOPE: Certificated to design, install and commission FM200 (pre-engineered/engineered) Argotec and CO_2 systems and issue LPCB Certificates of Conformity for these systems.

Surefire Systems Limited
Marston Court, 98-106 Manor Road, Wallington, Surrey SM6 0DW

Tel: +44 (0)181 773 3770 • Fax: +44 (0)181 669 6652

Certificate No. CFSI-001

SCOPE: Certificated to design, install and commission FM200 (pre-engineered/engineered), Argonite (Engineered) Halon and CO_2 systems and, issue LPCB Certificates of Conformity for these systems.

Wormald Ansul (UK) Limited, Wormald Fire Systems
Wormald Park, Grimshaw Lane, Newton Heath, Manchester M40 2WL

Tel: +44 (0)161 205 2321 • Fax: +44 (0)161 455 4459

Certificate No. CFSI-002

SCOPE: Certificated to design, install and commission Inergen systems and issue LPCB Certificates of Conformity for these systems.

INTRODUCTION

Equipment has been assessed and tested to the appropriate Standard, and the firms' manufacturing facilities have been certificated to ISO 9002 as a minimum. These facilities are re-inspected twice per annum and in addition the equipment may be subject to re-examination and testing.

LPCB certificated or approved equipment shall be used, where available.

In addition to approved equipment listed below the following schemes are now also operational:

Gaseous System Components.

Foam Systems.

Dry Powder Systems.

Dry Powder System Components.

Fine Water Spray Systems.

7 SECTION 2: APPROVED EQUIPMENT
2.1: GASEOUS FIXED FIRE FIGHTING SYSTEM DESIGN MANUALS

Introduction

Design Manual approval does not imply any component or system approval.

Approval Conditions here set out take precedence over any material contained in such Design Manuals for the purpose of LPCB recognition for life and property protection purposes.

Design Manuals are evaluated to appropriate Standards and cold discharge tests are conducted to validate design calculation methods[N1,2]. Recognition is given for compliance with Clauses 1-9 and Notes 1-6.

Design Manual Approval Conditions

1. All Design manual Approvals are subject to Gaseous System Approval (ref. Part 7, Section 2.3) prior to Dec. 99 for continued LPCB Design Manual Approval[N2]

2. Minimum design concentration by volume shall be the minimum extinguishing concentration for the risk enclosure, at the minimum ambient temperature liable to be experienced in the enclosure, plus at least 30% safety factor[N2,3]

3. The minimum design concentration for the risk shall take into account the latest state of recognised research knowledge, including the July 96 Loss Prevention Council Report titled 'Halon Alternatives: A report on the fire extinguishing performance characteristics of some gaseous alternatives to Halon 1301'[N2]

4. Design centres shall be covered by a certificated ISO 9001 quality system.

5. All designs shall be conducted by competent personnel trained in the correct use of the design manual and system software.

6. The system shall be designed in accordance with NFPA 2001 and the BFPSA "Code of Practice for Gaseous Fire Fighting Systems", but subject to compliance with any additional or different LPCB requirements. In due course the draft ISO standard "Gaseous Fire Extinguishing Systems" may also be involved.

7. Enclosure integrity shall be ensured, by the conduct of test(s), with satisfactory results.

8. The structural integrity and venting requirements for enclosures shall be taken into account and processed through a certificated ISO 9001 system at the design stage. This should include provision for both pressure relief during discharge and post fire venting in accordance with specified requirements.

9. It shall be ensured that appropriate safety measures are applied, taking into account BFPSA Code of Practice and the Halon Alternative Group report 'A review of the toxic and asphyxiating hazards of clean agent replacements for Halon 1301'.

Note 1: LPCB does not assure the suitability of agent selection, design concentration or method of application for a protected risk. The quantity and type of agent required for any risk will be dependent on many factors including materials within the protected enclosure, enclosure geometry, environmental conditions, potential fire conditions and combinations thereof. No guarantee as to the effectiveness or safety of any particular agent or design should be inferred, nor should anything herein be construed as a recommendation or authorisation. All usage of the information herein or related details is intended for use by technically competent personnel, who deploy such information at their own risk.

Note 2: LPC tests (Ref. LPR6: July 96) have shown minimum design concentrations for FM200, Argotec, Inergen and Argonite systems of 8.6%, 42%, 40% and 39% respectively were adequate to protect the described test risks. These concentrations are considered to be LPCB minimum levels for any fire type, subject to review on successful completion of gaseous system approval testing prior to Dec 99 for continued LPCB Design Manual approval.

Note 3: Design Manual technical data relating to extinguishing concentration and fire loads not in compliance with all these Clauses and Notes does not form part of this recognition and is not validated by the LPCB.

Note 4: Where possible, gaseous systems shall be installed by LPCB LPS 1204 certificated firms. Installation and commissioning practices are not otherwise covered by this approval.

Note 5: Service & maintenance practice is not covered by this approval.

Note 6: Design manuals approved by the LPCB are subject to on-going monitoring and evaluation processes.

Chubb Fire Limited

Lancaster Road, Cressex Business Park, High Wycombe, Bucks. HP12 3QF

Tel: +44 (0)1494 441114 • Fax: +44 (0)1494 531710

Chubb FM200 Engineered gaseous fire extinguishing systems design manual[1]

1. The system shall be designed in accordance with NFPA 2001 together with appropriate parts of BS 5306 Section 5.1 and Design Manual Approval Conditions specified in Part 7 of the LPCB List of Approved Fire and Security Products and Services.

Ginge-Kerr

526 Fleet Lane, St. Helens, Merseyside WA9 2NB

Tel: +44 (0)1744 28671 • Fax: +44 (0)1744 20174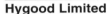

Ginge-Kerr Argonite engineered fire extinguishing system design manual[1]

1. The system shall be designed in accordance with the Design Manual Approval Conditions specified in Part 7 of the LPCB List of Approved Fire and Security Products and Services..

Hygood Limited

Woodlands Place, Woodlands Road, Guildford, Surrey GU1 1RN

Tel: +44 (0)1483 572222 • Fax: +44 (0)1483 302180

Hygood FM-200 Pre-engineered/Engineered gaseous fire extinguishing systems design manual[1]

1. The system shall be designed in accordance with NFPA 2001 together with appropriate parts of BS 5306 Section 5.1 and Design Manual Approval Conditions specified in Part 7 of the LPCB List of Approved Fire and Security Products and Services.

Kidde Fire Protection Limited

Belvue Road, Northolt, Middlesex UB5 5QW

Tel: +44 (0)181 845 7711 • Fax: +44 (0)181 845 4304

Kidde Fire Protection GX20 (FM200) Pre-engineered/Engineered gaseous fire extinguishing systems design manual[1]

1. The system shall be designed in accordance with NFPA 2001 together with appropriate parts of BS 5306 Section 5.1 and Design Manual Approval Conditions specified in Part 7 of the LPCB List of Approved Fire and Security Products and Services.

Preussag Fire Protection Limited

Field Way, Greenford, Middlesex UB6 8UZ

Tel: +44 (0)181 832 2000 • Fax: +44 (0)181 832 2200

Preussag Argotec Engineered gaseous fire fighting system design manual[1]

1. The system shall be designed in accordance with the Design Manual Approval Conditions specified in Part 7 of the LPCB List of Approved Fire and Security Products and Services..

Thorn Security Limited

160 Billett Road, Walthamstow, London E17 5DR

Tel: +44 (0)181 919 4000 • Fax: +44 (0)181 919 4040

Thorn FM200 Pre-engineered/Engineered gaseous fire fighting systems design manual[1]

1. The system shall be designed in accordance with NFPA 2001 together with appropriate parts of BS 5306 Section 5.1 and Design Manual Approval Conditions specified in Part 7 of the LPCB List of Approved Fire and Security Products and Services

Wormald Ansul (UK) Limited, Wormald Fire Systems

Wormald Park, Grimshaw Lane, Newton Heath, Manchester M40 2WL

Tel: +44 (0)161 205 2321 • Fax: +44 (0)161 455 4459

Wormald Inergen Pre engineered/Engineered gaseous fire extinguishing system design manual[1]

1. The system shall be designed in accordance with the Design Manual Approval Conditions specified in Part 7 of the LPCB List of Approved Fire and Security Products and Services.

7 SECTION 2:
2.2 GASEOUS SYSTEM COMPONENTS

INTRODUCTION

1. All LPCB listed gaseous system components are subject to LPCB on-going monitoring and evaluation processes.

2. Approved components are intended for use in conjunction with LPCB Approved Design Manuals.

Chubb Fire Limited

Lancaster Road, Cressex Business Park, High Wycombe, Bucks HP12 3QF

Tel: +44 (0)1494 441114 • Fax: +44 (0)1494 531710

CHUBB FM200 FIXED FIRE FIGHTING COMPONENTS

a   This range of equipment is suitable for storage temperatures from -20°C to +55°C and is approved for use with HFC 227ea (FM200) & Nitrogen.

b   This range of equipment is approved for use in conjunction with the Chubb Fire Ltd, LPCB approved, Design Manual that also relates to HFC 227ea (FM200).

c   System pressure is specified at 25 bar @ 20°C and max service pressure of 34.5 bar @ 55°C.

Description	Capacity (ltrs)	Part No.	Bracket No.
Container Valve, Container and Container Bracket Assemblies			
6 Ltr Container Assembly	6	4006[1,2,3]	40.817.11
10 Ltr Container Assembly	10	4010[1,2,3]	40.817.11
15 Ltr Container Assembly	15	4015[1,2,3]	40.817.11
25 Ltr Container Assembly	25	4025[1,2,3]	40.817.13
41 Ltr Container Assembly	41	4041[1,2,3]	40.817.13
63 Ltr Container Assembly	63	4063[1,2,3]	40.817.13
100 Ltr Container Assembly	100	4100[1,2,3]	40.817.15
155 Ltr Container Assembly	155	4155[1,2,3]	40.817.17
241 Ltr Container Assembly	241	4241[1,2,3]	40.817.17

1. These containers are designed and manufactured to BS5045 Part 2.
2. Available with and without contents monitoring pressure switch.
3. An extension to the part number gives the container fill.

Description	Part No.	
Container Valves		
CV21 Container Valve	49/90421	
CV45 Container Valve	49/90422	
CV60 Container Valve	49/90423	

Description	Part No.	Nominal Bore (mm)
Discharge Hoses		
Standard Discharge Hose	40.815.01	25
Discharge Hose C/W check valve	40.815.11	25
Standard Discharge Hose	40.815.02	40
Discharge Hose with St.St. ends	40.815.22	40
Standard Discharge Hose	40.815.03	50
Discharge Hose with St.St. ends	40.815.23	50
Standard Discharge Hose	40.815.04	65
Discharge Hose with St.St. ends	40.815.24	65

Continued

Chubb Fire Limited (continued)

Description	Part No.	Nominal Bore (mm)
Discharge Manifolds Including Check Valves and Brackets		
Two point manifold	40.816.28	80
Three point manifold	40.816.38	80
Two point manifold	40.816.21	100
Three point manifold	40.816.31	100
Main/Reserve Tee Check	40.816.05	50

Description	Part No.
Pressure Switches	
Container Pressure Switch	40/70355
Discharge Pressure Switch	40.818.01

Description	Part No.	Nominal Bore (mm)
Discharge Nozzles		
360° Nozzle	45410[4,5,6]	10
360° Nozzle	45415[4,5,6]	15
360° Nozzle	45420[4,5,6]	20
360° Nozzle	45425[4,5,6]	25
360° Nozzle	45432[4,5,6]	32
360° Nozzle	45440[4,5,6]	40
360° Nozzle	45450[4,5,6]	50
180° Nozzle	45210[4,5,6]	10
180° Nozzle	45215[4,5,6]	15
180° Nozzle	45220[4,5,6]	20
180° Nozzle	45225[4,5,6]	25
180° Nozzle	45232[4,5,6]	32
180° Nozzle	45240[4,5,6]	40
180° Nozzle	45250[4,5,6]	50

4. BSPT and NPT thread options available.
5. Material options include aluminium and brass.
6. Nozzles drilled in accordance with system calculations.

Description	Part No.
Valve Actuators	
Pneumatic Actuator	49.700.01
Solenoid Actuator	49.700.02
Manual Actuator	49.700.03
Pilot Vent Valve	49.700.05
Manual/Pneumatic Actuator	49.700.06

Description	Part No.	Nominal Bore
Pilot Hoses		
Pilot Hose 500mm	40/90181	3/16"
Pilot Hose 725mm	40/90180	3/16"

Fireater Limited

Fireater House, South Denes Road, Great Yarmouth, Norfolk NR30 3QP

Tel: +44 (0)1493 859822 • Fax: +44 (0)1493 858374

FIREATER FE-13 FIXED FIRE FIGHTING COMPONENTS

a)   This range of equipment is suitable for storage temperatures from -20° to +50°C.
b)   System pressure is specified as 41 bar @ 20°C and max. service pressure of 115 bar 50°C.

Description	Capacity	Part no.
Cylinder valve, cylinder and skid assemblies		
34L Cylinder Assembly	17 - 27.2	F2656[1]
40L Cylinder Assembly	20 - 32	F2655[1]
2-34L Cylinder Assembly - Mechanically operated	34 - 54.4	F2870[1]
2-34L Cylinder Assembly - Protractor operated	34 - 54.4	F2872[1]
2-34L Cylinder Assembly - Solenoid operated	34 - 54.4	F2874[1]
2-40L Cylinder Assembly - Mechanically operated	40 - 64	F2871[1]
2-40L Cylinder Assembly - Protractor operated	40 - 64	F2888[1]
2-40L Cylinder Assembly - Solenoid operated	40 - 64	F2873[1]
3-40L Cylinder Assembly - Mechanically operated	60 - 96	F2875[1]
3-40L Cylinder Assembly - Protractor operated	60 - 96	F2876[1]
3-40L Cylinder Assembly - Solenoid operated	60 - 96	F2889[1]
Solenoid Pilot Cylinder Assembly	0.4 litres	F1416[1]

1 These cylinders are designed and manufactured to BS 5045 Pt.1

Description	Part no.
Cylinder valve	
019 FE-13 Cylinder Valve	F2550
Manifold and outlet adaptors	
2-Way Check Valve Manifold Assembly	F2658
Swivel Connector Kit	F2830
Outlet Adaptor	F2561
Discharge nozzle	
æ" BSPT 360° Discharge Nozzle	F2659
Pressure switches	
Discharge Pressure Switch Assembly	F1505
Valve actuators	
Manual/Pneumatic Actuator	F1125
Manual/Electric Actuator	F1162[2]

2   The Manual/Electric Actuator is used with Protractor F1557.

Actuation equipment	
Pilot Vent Assemblies	F1247
Pull Box Assembly	F1274
Corner Pulley	F1272
2-Cylinder Lever Link Assembly	F3009
3-Cylinder Lever Link Assembly	F3008
Pull Cable	F1273
Protractor	F1557
Pilot Hose 280mm	F2884
Pilot Hose 375mm	F1091
Pilot Hose 510mm	F1092

Ginge-Kerr
526 Fleet Lane, St Helens, Merseyside, WA9 2NB

Tel: +44 (0)1744 28671 • Fax: +44 (0)1744 20174

GINGE-KERR ARGONITE FIXED FIRE FIGHTING COMPONENTS
1.This range of equipment is suitable for storage temperatures from -20°C to +50°C and is approved for use with Argonite.
2.This range of equipment is approved for use in conjunction with the Ginge-Kerr Ltd, LPCB approved Argonite Design Manual.
3.System pressure is specified at 150 bar and 200 bar @ 15°C.

Description	Capacity (ltr)	Part No.	System Pressure (bar)
Container and Container Bracket Assemblies			
67.5 Litre Cylinder	67.5	K4112	150[4]
80 Litre Cylinder	80	K4111	150[4]
67.5 Litre Cylinder	67.5	K4207	200[4]
80 Litre Cylinder	80	K4208	200[4]

4. These containers are designed and manufactured to BS 5045 Pt. 1
5. A range of wall brackets, clamping bars, spacers and clamping bolts for secure container installation are also available.

Discharge Manifolds Including Check Valves and Brackets

Description	Part No.	Nominal Bore (mm)	System Pressure (bar)
Discharge Manifold - Single Stub	K4174-02/14	25	150
Discharge Manifold - Double Stub	K4175-04/42	25,40,50	150
Discharge Manifold - Single Stub	K4199-02/06	25	200
Discharge Manifold - Double Stub	K4200-104/308	25,40,50	200
Manifold Non Return Valve	01-6451-0000	½" BSP	150/200

Description	Part No.	Orifice Size[7] (mm)	System Pressure (bar)
Restrictor Assembly			
Restrictor Assembly - ½" BSP	01-3454-1200	2-10	150/200
Restrictor Assembly - 1" BSP	01-3454-1300	7-19	150/200
Restrictor Assembly - 1½" BSP	01-3454-1400	12-28	150/200

6 Nozzles sized in accordance with system calculations

Description	Part No.	Nominal Bore (mm)	System Pressure (bar)
Diverter/Selector Valve and Equipment			
Pressure Operated Valve	K3864/8	20,25,32,40,50	150

Description	Part No.	Orifice Size[7] (mm)	System Pressure (bar)
Discharge Nozzles			
Argonite Nozzle Assembly - ½" BSPT	01-3460-1200	2-8	150/200
Argonite Nozzle Assembly - ¾ BSPT	01-3460-1300	7-12	150/200
Argonite Nozzle Assembly - 1" BSPT	01-3460-1400	10-16	150/200
Argonite Nozzle Assembly - 1½" BSPT	01-3460-1500	15-20	150/200

7 Nozzle Size in accordance with system calculation

Description	Part No.	Nominal Bore (mm)	System Pressure (bar)
Discharge Hoses			
Discharge Hi-Flex Hose	01-3261-0100	⅜"x 400	150
Discharge Hi-Flex Hose	01-3281-0100	⅜" x 400	200
Actuator Hi-Flex Hose	03-3272-01/0200	¼" x 250/500	150
Actuator Hi-Flex Hose	01-3272-01/0200	¼"x 250/500	200

Continued

Ginge-Kerr (continued)

Description	Part No.	System Pressure (bar)
Container Valves		
Pneumatic Argonite Valve	01-6410-0100	150
Pneumatic Argonite Valve	01-6410-0200	200
Solenoid Valve, Pressure Switches and Gauges		
Solenoid Valve with Pressure Gauge & Switch	01-7172-0100	150
Pressure Gauge and Switch	01-7171-0100	150
Pressure Gauge	01-7174-0100	150/200
Solenoid Valve with Pressure Gauge & Switch	01-7172-0200	200
Pressure Gauge and Switch	01-7171-0200	200
Pressure Switches		
Pressure Operated Switch	03-5713-0000	150/200
Valve Actuators		
Pneumatic Actuator	03-4161-0000	150/200
Manual/Pneumatic Actuator	03-4162-0000	150/200

Hygood Limited

Woodlands Place, Woodlands Road, Guildford, Surrey GU1 1RN

Tel: +44 (0)1483 572222 • Fax: +44 (0)1483 302180

HYGOOD FM200 FIXED FIRE FIGHTING COMPONENTS

1. This range of equipment is suitable for storage temperatures from 0°C to 50°C and is approved for use with FM200®.
2. This range of equipment is approved for use in conjunction with Hygood Ltd, LPCB approved FM200 Installation Guide.
3. System pressure is specified at 25 bar at 20°C.

Description	Part No.	Capacity (litres)
Valve and Container Assembly		
Container Assembly	9055	6.5
Container Assembly	9060	13
Container Assembly	9065	25
Container Assembly	9071	55
Container Assembly	9075	106
Container Assembly	9080	147
Container Assembly	9085	180

4. Containers are designed and manufactured in accordance with BS5045 Part 2 Class 1A.
5. Containers are designed to be mounted vertically.

Description	Part No.
Container Bracket Assembly	
6.5-25 litre Container	9000
55 litre Container	9010
106-180 litre Container	9012
Valve Actuators	
Solenoid Actuator 24v dc BASEEF A	95530
Solenoid Actuator 24v dc	9400
Pneumatic Actuator	2900
Manual Actuator	2880

Continued

Hygood Limited (continued)

Description	Part No.	Nominal bore (mm)
Discharge Hoses & Pilot Valve		
50mm Discharge Hose	6540	50
8mm Pilot Hose x 710mm	6490	8
Discharge Nozzles		
10mm Aluminium Nozzle	3380	10
15mm Aluminium Nozzle	3390	15
20mm Aluminium Nozzle	3400	20
25mm Aluminium Nozzle	3410	25
32mm Aluminium Nozzle	3420	32
40mm Aluminium Nozzle	3430	40
50mm Aluminium Nozzle	3440	50

6. 360° and 180° discharge patterns available.
7. BSPP thread types.
8. Nozzles drilled in accordance with system calculation.

Description	Part No.	Nominal bore (mm)
Manifold & CV Assembly		
2 Port 65mm Manifold & Check Valve	9335	
3 Port 65mm Manifold & Check Valve	9336	
4 Port 65mm Manifold & Check Valve	9337	
2 Port 80mm Manifold & Check Valve	9352	
3 Port 80mm Manifold & Check Valve	9353	
4 Port 80mm Manifold & Check Valve	9345	
5 Port 80mm Manifold & Check Valve	9346	
2 Port 100mm Manifold & Check Valve	9355	
3 Port 100mm Manifold & Check Valve	9356	
4 Port 100mm Manifold & Check Valve	9354	
5 Port 100mm Manifold & Check Valve	9357	
6 Port 100mm Manifold & Check Valve	9358	
3 Port 150mm Manifold & Check Valve	9312	
4 Port 150mm Manifold & Check Valve	9313	
5 Port 150mm Manifold & Check Valve	9314	
6 Port 150mm Manifold & Check Valve	9315	
7 Port 150mm Manifold & Check Valve	9316	
8 Port 150mm Manifold & Check Valve	9317	
9 Port 150mm Manifold & Check Valve	9318	
10 Port 150mm Manifold & Check Valve	9319	
Manifold Bracket Assembly		
65mm Nominal Bore Manifold	30460	
80mm Nominal Bore Manifold	30500	
100mm Nominal Bore Manifold	30510	
150mm Nominal Bore Manifold	30515	
Manifold Burst Disc Assembly		
Manifold Burst Disc Assembly	2790	
Pressure Switches		
Discharge Pressure Switch	99210	
Watchman Weigh Base		
425mm x 425mm Platform	9370	
300mm x 300mm Platform	9360	
Warning Notices		
Door Warning Notice NOAEL	2676	
Door Warning NoticeNOAEL	2675	
Manual Release Notice	2659	

Kidde Fire Protection Limited
Belvue Road, Northolt, Middlesex, UB5 5QW

Tel: +44 (0)181 845 7711 • Fax: +44 (0)181 845 4304

KIDDE FM200 FIXED FIRE FIGHTING COMPONENTS

a. This range of equipment is suitable for storage temperatures from -20°C to +54°C and is approved for use with FM200.
b. This range of equipment is approved for use in conjunction with the Kidde Fire Protection Ltd, LPCB approved FM200 Design Manual.
c. System pressure is specified at 25 bar @ 20°C and max service pressure of 34.5 bar @ 55°C.
d. Nitrogen pilot pressure is specified as 124 bar @ 20°C.

Description	Capacity (ltr)	Part No.	Strap	Bracket No.
Container Valve, Container and Container Bracket Assemblies				
Cyl & GCV40 Valve Assy	$5^{1,2}$	E7763-101	C6331-103	D1162-001
Cyl & GCV40 Valve Assy	$5^{1,3}$	E7763-121	C6331-103	D1162-001
Cyl & GCV40 Valve Assy	$8^{1,2}$	E7763-102	C6331-103	D1162-002
Cyl & GCV40 Valve Assy	$8^{1,3}$	E7763-122	C6331-103	D1162-002
Cyl & GCV40 Valve Assy	$16^{1,2}$	E7763-103	C6331-104	D1162-003
Cyl & GCV40 Valve Assy	$16^{1,3}$	E7763-123	C6331-104	D1162-003
Cyl & GCV40 Valve Assy	$28^{1,2}$	E7763-104	C6331-104	D1162-004
Cyl & GCV40 Valve Assy	$28^{1,3}$	E7763-124	C6331-104	D1162-004
Cyl & GCV50 Valve Assy	$51^{1,2}$	E7763-105	C6331-105	
Cyl & GCV50 Valve Assy	$81^{1,2}$	E7763-106	C6331-105	
Cyl & GCV50 Valve Assy	$142^{1,2}$	E7763-108	C6331 -106	
Cyl & GCV65 Valve Assy	$243^{1,2}$	E7763-110	C6331-107	
Nitrogen Pilot Cylinder	1.77	B6793-720		B6793-721

1.Cylinders are designed and manufactured to US CFR49: DOT 4BW 500 (HSE)
2.Vertical mount cylinder with straight syphon tube
3.Horizontal mount cylinder with angled syphon tube

Description	Part No.	
Container Valves		
GCV40 valve	K63536-03	
GCV50 valve	K63536-02	
GCV65 valve	K63536-01	

Description	Part No.	Nominal Bore (mm)
Discharge Hoses & Outlet Adaptors		
GCV40 Outlet Hose	B6793-760	40
GCV50 Outlet Hose	B6793-761	50
GCV65 Outlet Hose	B6793-763	65
GCV40 Outlet Adaptor	K63497-02	40
GCV50 Outlet Adaptor	K63497-03	50
GCV65 Outlet Adaptor	K63497-04	65
Discharge Manifolds Including Check Valves and Brackets		
Elbow Check	B6793-762	50
Elbow Check	B6793-764	65
2 Port Manifold	D1511-006	50
2 Port Manifold	D1511-007	80
2 Port Manifold	D1511-009	100
3 Port Manifold	D1511-008	80
3 Port Manifold	D1511-010	100

Continued

Kidde Fire Protection Limited (continued)

Description	Part No.
Pressure Switches & Devices	
Cylinder Contents Supervisory Pressure Switch	B6793-730
Discharge Pressure Switch	B6793-731
Discharge Pressure Switch (Explosion Proof)	B6793-732
Pressure Trip	B6793-733

Description	Part No.	Nominal Bore (mm)
Discharge Nozzles		
180° Nozzle[4,5]	C3333-301-xx	15
180° Nozzle[4,5]	C3333-302-xx	20
180° Nozzle[4,5]	C3333-303-xx	25
180° Nozzle[4,5]	C3333-304-xx	32
180° Nozzle[4,5]	C3333-305-xx	40
180° Nozzle[4,5]	C3333-306-xx	50
360° Nozzle[4,5]	C3333-307-xx	15
360° Nozzle[4,5]	C3333-308-xx	20
360° Nozzle[4,5]	C3333-309-xx	25
360° Nozzle[4,5]	C3333-310-xx	32
360° Nozzle[4,5]	C3333-311-xx	40
360° Nozzle[4,5]	C3333-312-xx	50

4. Nozzles drilled in accordance with system calculations.
5. Nozzles are manufactured from brass.

Description	Part No.
Valve Actuators	
Electric Control Head	B6793-701
Electric & Cable Control Head	B6793-702
Cable & Electric Control Head (Flameproof)	B6793-703
Cable Operated Control Head	B6793-704
Lever Operated Control Head	B6793-705
Pressure & Lever Operated Control Head	B6793-706
Pressure Operated Control Head	B6793-707
Pressure Operated Control Head[6]	B6793-708
Electric Control Head[6]	B6793-709

6. Stackable configuration

Pilot Equipment

Master Cylinder Adaptor Kit	B6793-770
559mm Long Pilot Hose	B6793-771
762mm Long Pilot Hose	B6793-772
Male Straight Adaptor	B6793-773
Male Elbow Adaptor	B6793-774
Male Branch Tee	B6793-775

Thorn Security Limited
160 Billet Road, Walthamstow, London E17 5DR

Tel: +44 (0)181 919 4000 • Fax: +44 (0)181 919 4040

THORN SECURITY FM200 FIXED FIRE FIGHTING COMPONENTS

1. This range of equipment is suitable for storage temperatures from 0°C to 50°C and is approved for use with FM200®.
2. This range of equipment is approved for use in conjunction with Thorn Security, LPCB approved FM200 Design Manual.
3. System pressure is specified at 25 bar at 20°C.

Description	Part No.	Capacity (litres)
Valve and Container Assembly		
6.5l Container Assembly	303-205-001	6.5
13l Container Assembly	303-205-002	13
25.5l Container Assembly	303-205-003	25
52l Container Assembly	303-205-004	52
106l Container Assembly	303-205-005	106
147l Container Assembly	303-205-006	147
180l Container Assembly	303-205-007	180

4. Containers are designed and manufactured in accordance with BS5045 Part 2 Clause 1A.
5. Containers are designed to be mounted vertically.

Description	Part No.
Container Bracket Assembly	
6.5l Container	311-205-001
13l Container	311-205-002
25.5l Container	311-205-003
52-180l Container	311-205-004
Valve Actuators	
Solenoid Actuator 24v dc BASEEFA	304-205-002
Solenoid Actuator 24v dc	304-205-001
Pneumatic Actuator	304-205-004
Manual Actuator	304-205-003

Description	Part No.	Nominal bore (mm)
Discharge Hoses & Pilot Valve		
50mm Discharge Hose	306-205-002	50
Discharge Nozzles		
10mm Aluminium Nozzle	310-205-019	10
15mm Aluminium Nozzle	310-205-007	15
20mm Aluminium Nozzle	310-205-020	20
25mm Aluminium Nozzle	310-205-008	25
32mm Aluminium Nozzle	310-205-021	32
40mm Aluminium Nozzle	310-205-009	40
50mm Aluminium Nozzle	310-205-010	50

8. 360° and 180° discharge patterns available.
9. Nozzles drilled in accordance with system calculation.

Continued

Thorn Security Limited (continued)

Description	Part No.
Manifold & CV Assembly	
2 Port 65mm Manifold & Check Valve	TSL9335
3 Port 65mm Manifold & Check Valve	TSL9336
4 Port 65mm Manifold & Check Valve	TSL9337
2 Port 80mm Manifold & Check Valve	307-205-001
3 Port 80mm Manifold & Check Valve	307-205-002
4 Port 80mm Manifold & Check Valve	307-205-003
2 Port 100mm Manifold & Check Valve	307-205-004
3 Port 100mm Manifold & Check Valve	307-205-005
4 Port 100mm Manifold & Check Valve	307-205-006
5 Port 100mm Manifold & Check Valve	307-205-007
6 Port 100mm Manifold & Check Valve	307-205-008
3 Port 150mm Manifold & Check Valve	307-205-009
4 Port 150mm Manifold & Check Valve	307-205-010
5 Port 150mm Manifold & Check Valve	TSL9314
6 Port 150mm Manifold & Check Valve	TSL9315
7 Port 150mm Manifold & Check Valve	TSL9316
8 Port 150mm Manifold & Check Valve	TSL9317
9 Port 150mm Manifold & Check Valve	TSL9318
10 Port 150mm Manifold & Check Valve	TSL9319
Manifold Bracket Assembly	
65mm Nominal Bore Manifold	TSL30460
80mm Nominal Bore Manifold	TSL30500
100mm Nominal Bore Manifold	TSL30510
150mm Nominal Bore Manifold	TSL30515
Manifold Burst Disc Assembly	
Manifold Burst Disc Assembly	TSL2790
Pressure Switches	
Discharge Pressure Switch	305-205-002
Container Load Cell	
425mm x 425mm Platform	TSL9370
300mm x 300mm Platform	TSL9360
Warning Notices	
Door Warning Notice <NOAEL	314-205-002
Door Warning Notice >NOAEL	314-205-001
Manual Release Notice	314-205-003

GASEOUS SYSTEM GENERAL APPROVAL CONDITIONS

To obtain Gaseous System approval firms shall satisfy LPCB in respect of the following:

a) The related Design Manual shall be approved and listed.

b) Component evaluation including component compatibility will have been conducted by the LPCB with the firm's components also being listed in the Section Gaseous System Components.

c) The firm shall provide evidence to LPCB in respect of the capability of the system to extinguish fires in representative fire scenarios, with LPCB system approval verification fire testing as necessary.

Note: Such systems are only recognised when they are installed by LPCB LPS 1204 certificated firms.

Recommendation to Users/Specifiers/Insurers/Interested Parties

It is recommended that the responsibility of the supplier for provision of the system in total, (including the integration of the following facets), be established at the contract stage:

• Risk assessment.

• Extinguishing design concentration.

• Enclosure structural/leakage integrity.

• Discharge/Post discharge venting.

• Electrical system integration.

Central Sprinkler Corporation

Corringham Road Industrial Estate, Gainsborough, Lincolnshire DN21 1QB

Tel: +44 (0)1427 615401 • Fax: +44 (0)1427 610433
Sales enquiries: Tel: +1 215 362 0700 • Fax: +1 215 362 5385

Central Model SS07FSP Open Foam Sprayer [1,2]

1. To be used in accordance with the manufacturer's instructions, titled "Caution, service, maintenance and installation instructions for Central SS07F open foam sprayer".
2. Application is limited to where the purchaser, authority having jurisdiction and/or insurer consent to use.

Spraysafe Automatic Sprinklers Limited

Corringham Road Industrial Estate, Gainsborough, Lincolnshire DN21 1QB

Tel: +44 (0)1427 615401 • Fax: +44 (0)1427 610433

Spraysafe Model SS07F Open Foam Sprayer [1,2]

1. To be used in accordance with the manufacturer's instructions, titled "Caution, service, maintenance and installation instructions for Spraysafe SS07F open foam sprayer".
2. Application is limited to where the purchaser, authority having jurisdiction and/or insurer consent to use.

7 SECTION 4:

4.1: FINE WATER SPRAY COMPONENTS

Angus Fire Armour Limited

Kastanievej 15, DK-5620, Glamsbjerg, Denmark

UK Sales Enquiries: Tel: +44 (0)1844 214545 • Fax: +44 (0)1844 213511

† ANGUS LOFLOW K11 Fine Water Spray Sprinkler [1,2]

†   Indicates glass bulb type

1.   To be installed and used only in accordance with the manufacturer's instructions.
2.   Applications are limited to where the purchaser, authority having jurisdiction and/or insurer consent to use.

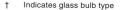

G.W. Sprinkler A/S

Kastanievej 15, DK - 5620 Glamsbjerg, Denmark

Tel: +45 64 72 2055 • Fax: +45 64 72 2255

† GW LOFLOW K11 Fine Water Spray Sprinkler [1,2]

†   Indicates glass bulb type

1.   To be installed and used only in accordance with the manufacturer's instructions.
2.   Applications are limited to where the purchaser, authority having jurisdiction and/or insurer consent to use.

INTRODUCTION

The LPCB Approval Scheme for fixed fire extinguishing systems for catering equipment has been developed in recognition of the specific risk posed by oil fires in cooking appliances of catering or food production establishments.

The objective of the scheme is to ensure that the Listed systems provide adequate protection of the cooking appliances, hood, filters and extract system when correctly installed, serviced and maintained by trained personnel.

This LPCB Approval Scheme combines two main elements in its approval of fixed extinguishing systems:

i.   system assessment to LPS 1223: *Requirements and Testing Procedures for Approval of Fixed Fire Extinguishing Systems for Catering Equipment*; and

ii. assessment of installer competence through LPCB review of the Listed company.

LPS 1223: *Requirements and Testing Procedures for Approval of Fixed Fire Extinguishing Systems for Catering Equipment*

LPS 1223 specifies the requirements and test procedures for assessment of fixed fire extinguishing systems designed for the protection of catering equipment, such as deep fat fryers*, other cooking appliances and associated extract systems, from fires originating in the cooking appliance.

The main requirements of LPS 1223 are:

• system specification

• design manual assessment

• component testing

• fire extinguishment tests

• system testing, including extinguishing agent discharge tests

Traditional extinguishing agents do not readily extinguish catering equipment oil fires because they do not provide sufficient cooling effect to reduce the cooking oil to below its auto-ignition temperature. The fire extinguishing tests in LPS 1223 recognise this requirement together with the need to ensure that neither the composition of the agent nor the manner in which it is discharged onto the fire causes flaming oil to be ejected.

Fish and chip ranges*

Fish and chip ranges are a specialised type of catering equipment that require special consideration with regards to protection due ostensibly to the position and orientation of the extract system.

Unless otherwise stated in the individual product entries, the Listed systems are not approved for protection of fish and chip ranges.

LPCB review of Listed companies

The LPCB reviews the Listed companies regularly to assess their continuing ability to meet the responsibilities detailed below:

- manufacture of the system and its components to the LPCB approved specification
- training of installers
- review of system designs and calculations
- review of installer competence
- issue of continuation of approval for individual installations on behalf of the LPCB

Approval of installed systems

Systems installed by trained personnel under the control of the Listed company shall be marked to indicate approval to LPS 1223. The initial period of approval for an individual system will be related to the service and maintenance requirements specified in the system manual and would normally be six months.

Continuation of approval is dependent on the system being serviced and maintained in accordance with the system manual. Servicing requirements include a review of the cleanliness of the extract system. This is considered to be an extremely important factor in reducing the risk of fire and system approval will not be reissued if other than light grease deposits are evident.

Amerex Corporation†

PO Box 81, Trussville, AL 35173-0081, USA

Tel: +1 205 655 3271 • Fax: +1 205 655 3279 • e-mail: sales@amerex-fire.com

UK Sales Enquiries: Amerex Fire International Limited
Tel: +44 (0)1633 87502 1 • Fax: +44 (0)1633 863595 • e-mail: amerexuk@aol.com

Approved Products
Amerex KP Fire Suppression System for catering equipment

LPCB Ref. No.
407a/01

Systems shall be designed, installed and maintained in accordance with Amerex Design, Installation, Maintenance and Recharge Manual Part No. 12385 Issued 1 May 1995 including Addenda 1 to 7.

† This company has not been assessed by the LPCB to ISO 9000

Part

8

SMOKE AND FIRE VENTILATION SYSTEMS

INTRODUCTION

This Part contains building products that are approved by the LPCB and were commercially available at the time of issue of this document.

Certificated products comply with the relevant LPCB technical requirements. Certification is maintained by product audits and ISO 9001 or 2 surveillances undertaken by product experts. Listing of a product signifies that it complies with the requirements of the LPC *Design Guide for the Fire Protection of Buildings.*

The use of such products, in conjunction with the aforementioned document, can have an effect on the building insurance premium thus potentially reducing the service costs of buildings.

CERTIFICATION

Requirements for LPCB Product Certification and Product Approval are defined in ISO 9000 and the appropriate Loss Prevention Standards from the following:

LPS 1162 *Requirements and Tests for Fire Dampers*

BS476: Part 24 *Fire tests on building materials and structures. Method for determination of the fire resistance of ventilation ducts.*

LPS 1182 *Requirements and tests for fire and smoke curtains.*

prEN12101-3 Fixed fire fighting systems, smoke and heat control systems, Part 3: Specification for powered smoke and heat exhaust ventilators.

Advanced Air (UK) Limited

3 & 4 Cavendish Road, Bury St. Edmunds, Suffolk IP33 3TE

Tel: +44 (0)1284 701351 • Fax: +44 (0)1284 701357

Certificate. No. 128a to LPS 1162

LPCB Ref No.	Name	Series	Maximum Duct Size			
			Oval, square, rectangular		Circular	
			Height mm	Width mm	Diameter mm	
128a/01	Curtain Fire	0100	1000	1000	1000	
	Damper	0200	1000	1000	1000	
128a/02	Fire Smoke	1240	1000	1000	1000	
	Control Damper	1250	1000	1000	1000	
		1260	1000	1000	1000	
		1270	1000	1000	1000	
128a/03	Insulated Fire Damper	2400	1000	1000	1000	

LPCB Ref No.	Fire Resistance Integrity (min)	Insulation (min)	Durability Tests and Corrosion Tests		
			Standard*	Marine †	Cyclic ‡
128a/01	Vertical plane 240	0	Satisfactory	Not tested	Not tested
	Horizontal plane 240	0			
128a/02	Vertical plane 120	0	Satisfactory	Not tested	Not tested
	Horizontal plane 120	0			
128a/03	Vertical plane 210	205	Satisfactory	Not tested	Not tested
	Horizontal plane 250	250			

Note: * Standard LPS 1162 requirement for normal use.
† Corrosion test to Appendix B of LPS 1162 for marine use.
‡ Cyclic test to Appendix A of LPS 1162 for repeated use.

Baxi Air Management Limited

South Street, Whitstable, Kent, CT5 3DU

Tel: +44 (0)1227 276100 • Fax: +44 (0)1227 264262

Certificate No. 017a to LPS 1162

LPCB Ref. No.	Trade Name	Series	Maximum Duct Size	Fire Resistance Integrity (min)	Durability Corrosion Tests		
					Standard*	Marine†	Cyclic‡
017a/01	Fire/Shield	101	1250mm x 1016mm	240 vertical	Satisfactory	Not tested	Not tested
		201	1250mm x 1016mm	240 horizontal	Satisfactory	Not tested	Not tested
		301	1016mm diameter		Satisfactory	Not tested	Not tested
		401	1250mm x 1016mm		Satisfactory	Not tested	Not tested
017a/02	Smoke/Shield	501	1016mm x 1016mm	120 vertical	Satisfactory	Not tested	Not tested
		601	1016mm diameter	120 horizontal	Satisfactory	Not tested	Not tested
		701	1016mm x 711mm		Satisfactory	Not tested	Not tested
017a/03	Trans/Shield	Sizes 1-9	450mm x 430mm	120 vertical	Satisfactory	Not tested	Not tested

Notes:
+ Smoke/Shield is known as Vent/Shield when used as a smoke vent in an evacuation system, both Smoke/Shield and Vent/Shield can be controlled by Actionpac - Multiplex control System.
* Standard LPS 1162 requirement for normal use.
† Corrosion test Appendix B LPS 1162 for marine use.
‡ Cyclic test Appendix A LPS 1162 for repeated use.

Cape Durasteel Limited

Bradfield Road, Wellingborough, Northamptonshire NN8 4HB

Tel: +44 (0)1933 440055 • Fax: +44 (0)1933 440066

Approval No. 008a to BS 476: Part 24

LPCB Ref. No.	Trade Name	BS 476: Part 24
008a/01	Duraduct	120 min

Fire Protection Limited

Millars 3, Southmill Road, Bishops Stortford, Hertfordshire CM23 3DH

Tel: +44 (0)1279 467077 • Fax: +44 (0)1279 466994

Approval No. 277a to BS 476: Part 24

LPCB Ref. No.	Trade Name	Duct Type	Condition	Minimum Wall Thickness	Type of Insulation	Insulation Thickness	Fire Resistance Insulation min	Integrity min
277a/01	Flamebar BW11	Ventilation	Fire outside	0.9mm	-	-	120	15
				1.2mm	-	-	240	15
				0.9mm	60kg/m^3	50mm	120	118
				0.9mm	105kg/m^3	50mm	120	120
				1.2mm	105kg/m^3	50mm	240	208
				1.2mm	105kg/m^3	120mm	240	240
		Ventilation	Fire inside	0.9mm	-	-	120	-
				1.2mm	-	-	240	-
				0.9mm	60kg/m^3	50mm	120	45
				0.9mm	105kg/m^3	50mm	120	60
				0.9mm	105kg/m^3	120mm	120	120
				1.2mm	105kg/m^3	120mm	240	120
		Smoke outlet	Fire outside	1.2mm	-	-	120	15
				1.6mm	-	-	240	15
				1.2mm	60kg/m^3	50mm	120	118
				1.2mm	105kg/m^3	50mm	120	120
				1.6mm	105kg/m^3	50mm	240	208
				1.6mm	105kg/m^3	120mm	240	240
		Smoke outlet	Fire inside	1.2mm	-	-	120	-
				1.6mm	-	-	240	-
				1.2mm	60kg/m^3	50mm	120	45
				1.2mm	105kg/m^3	50mm	120	60
				1.2mm	105kg/m^3	120mm	120	120
				1.6mm	105kg/m^3	120mm	240	120
		Kitchen extract	Fire outside	0.9mm	-	-	120	15
				1.2mm	-	-	240	15
				0.9mm	60kg/m^3	50mm	120	28*
				0.9mm	105kg/m^3	50mm	120	44*
				1.2mm	105kg/m^3	50mm	240	44*
				1.2mm	105kg/m^3	120mm	240	60*
		Kitchen extract	Fire inside	0.9mm	-	-	120	-
				1.2mm	-	-	240	-
				0.9mm	60kg/m^3	50mm	120	45
				0.9mm	105kg/m^3	50mm	120	60
				0.9mm	105kg/m^3	120mm	120	120
				1.2mm	105kg/m^3	120mm	240	120

Notes: 1 All ducts are protected with 0.7mm to 1.0mm of Flamebar BW11 Fire Proofing.
2 Maximum hanger centres are 1500mm for fire ratings up to two hours and 1000mm for fire ratings up to four hours.
3 Details for kitchen extract ducts also apply to ducts with combustible linings.
4 Insulation ratings marked with an asterisk refer to the internal face inside the furnace.

Firespray International Limited
Millars 3, Southmill Road, Bishops Stortford, Hertfordshire CM23 3DH

Tel: +44 (0)1279 467077 • Fax: +44 (0)1279 466994

LPCB

Certificate No. 415a to BS 476: Part 24

LPCB Ref. No.	Trade Name	Duct Type	Condition	Minimum Wall Thickness	Type of Insulation	Insulation Thickness	Fire Resistance Insulation min	Fire Resistance Integrity min
415a/01	Flamebar BW11	Ventilation	Fire outside	0.9mm	-	-	120	15
				1.2mm	-	-	240	15
				0.9mm	60kg/m³	50mm	120	118
				0.9mm	105kg/m³	50mm	120	120
				1.2mm	105kg/m³	50mm	240	208
				1.2mm	105kg/m³	120mm	240	240
		Ventilation	Fire inside	0.9mm	-	-	120	-
				1.2mm	-	-	240	-
				0.9mm	60kg/m³	50mm	120	45
				0.9mm	105kg/m³	50mm	120	60
				0.9mm	105kg/m³	120mm	120	120
				1.2mm	105kg/m³	120mm	240	120
		Smoke outlet	Fire outside	1.2mm	-	-	120	15
				1.6mm	-	-	240	15
				1.2mm	60kg/m³	50mm	120	118
				1.2mm	105kg/m³	50mm	120	120
				1.6mm	105kg/m³	50mm	240	208
				1.6mm	105kg/m³	120mm	240	240
		Smoke outlet	Fire inside	1.2mm	-	-	120	-
				1.6mm	-	-	240	-
				1.2mm	60kg/m³	50mm	120	45
				1.2mm	105kg/m³	50mm	120	60
				1.2mm	105kg/m³	120mm	120	120
				1.6mm	105kg/m³	120mm	240	120
		Kitchen extract	Fire outside	0.9mm	-	-	120	15
				1.2mm	-	-	240	15
				0.9mm	60kg/m³	50mm	120	28*
				0.9mm	105kg/m³	50mm	120	44*
				1.2mm	105kg/m³	50mm	240	44*
				1.2mm	105kg/m³	120mm	240	60*
		Kitchen extract	Fire inside	0.9mm	-	-	120	-
				1.2mm	-	-	240	-
				0.9mm	60kg/m³	50mm	120	45
				0.9mm	105kg/m³	50mm	120	60
				0.9mm	105kg/m³	120mm	120	120
				1.2mm	105kg/m³	120mm	240	120

Notes:
1. All ducts are protected with 0.7mm to 1.0mm of Flamebar BW11 Fire Proofing.
2. Maximum hanger centres are 1500mm for fire ratings up to two hours and 1000mm for fire ratings up to four hours.
3. Details for kitchen extract ducts also apply to ducts with combustible linings.
4. Insulation ratings marked with an asterisk refer to the internal face inside the furnace.

Guthrie Douglas Limited

12 Heathcote Way, Heathcote Industrial Estate, Warwick CV34 6TE

Tel: +44 (0)1926 452452 • Fax: +44 (0)1926 336417

Certificate No. 193a to LPS 1182: Issue 2.

LPCB Ref. No.	Trade Name	Curtain Type	Maximum Size		Grade
			Drape (m)	Width (m)	
193a/01	S.A.C. 100 Smoke Arrest Curtain	Fabric Smoke Curtain	2.9	6	A

Woods Air Movement Ltd
Tufnell Way, Colchester, Essex CO4 5AR

Tel: +44 (0)1206 544122 • Fax: +44 (0)1206 574434

Certificate No. 386a to prEN12101-3

LPCB Ref. No.	Trade Name	Temperature/ time classification		EN classification	Diameter mm	Blade material	Volume flow rate (indication)
386a/01	JM Aerofoil	200°C	120 min	F200	315 - 1600	Aluminium	up to 65m³/s
	HT Series	300°C	60 min	F300	315 - 1600	Aluminium	up to 65m³/s
		400°C	120 min	F400	315 - 1250	Aluminium	up to 30m³/s
		400°C	120 min	F400	630 - 1600	Steel	up to 35m³/s
386a/02	JM Aerofoil	200°C	120 min	F200	400 -1250	Aluminium	-
	bifurcated HT	300°C	60 min	F300	400 -1250	Aluminium	-
	series fans	400°C	120 min	F400	400 - 1250	Aluminium	-
		400°C	120 min	F400	400 - 1250	Steel	-
		600°C	90 min	F600	400 - 1250	Steel	up to 25m³/s
386a/03	DVA roof unit*	300°C	30 min	-	315 - 800	Aluminium	-
386a/04	UDA roof unit*	300°C	30 min	-	315 to 1600	Aluminium	-

* Note: Although cowls are not specifically addressed in prEN12101-3 the fan's have been tested in accordance with prEN12101-3 and the fans with cowls to an adhoc test for 30 mins.